GENETICS & SOCIETY

Edited by PENELOPE BARKER

THE REFERENCE SHELF

Volume 67 Number 3

THE H. W. WILSON COMPANY

New York 1995

THE REFERENCE SHELF

The books in this series contain reprints of articles, excerpts from books, and addresses on current issues and social trends in the United States and other countries. There are six separately bound numbers in each volume, all of which are generally published in the same calendar year. One number is a collection of recent speeches; each of the others is devoted to a single subject and gives background information and discussion from various points of view, concluding with a comprehensive bibliography that contains books and pamphlets and abstracts of additional articles on the subject. Books in the series may be purchased individually or on subscription.

Library of Congress Cataloging-in-Publication Data

Genetics & society / edited by Penelope Barker.
p. cm. — (The Reference shelf ; v. 67 no. 3)
Includes bibliographical references and index.
ISBN 0-8242-0870-6
1. Genetics—Social aspects. 2. Human genetics—Social aspects. 3. Genetic engineering—Social aspects. I. Barker, Penelope. II. Series.
QH438.7.G44 1995
306.4'61—dc20 95-10332
CIP

Cover: Two doctors inspect gene x-ray for disease.

Photo: AP/Wide World Photos

Printed in the United States of America

CONTENTS

III. Genetics and Industry

PREFACE

In 1953 two British scientists, then a 24-year-old graduate student, James D. Watson, and a 35-year-old physicist turned molecular biologist, Francis Crick, elucidated the structure of deoxyribonucleic acid, or DNA. Shaped like a twisted ladder, this "most golden of all molecules," as Watson termed it, had not long before been identified as the substance of which genes were made. Yet this discovery, which earned both Watson and Crick a Nobel Prize, and the technologies that were later developed to manipulate DNA pale in comparison to the advances made in the past decade, which are revolutionizing society in ways that few could have imagined just four decades ago.

Thanks to advances in genetic research, scientists have succeeded in tracking down a number of prominent disease genes, including the ones that cause Huntington's disease (HD), which claimed the life of the singer Woodie Guthrie, cystic fibrosis (CF), the most common genetic disease among Caucasians, and several kinds of cancer. Once a disease gene has been isolated, tests can be developed for screening purposes, enabling, for example, prospective parents with a family history of a fatal genetic disease to take a simple test to find out what the odds are that they will have a healthy child. A pregnant woman, too, can be tested, so she can make an informed decision about whether or not to continue the pregnancy. Or an individual who finds from such a test that he or she is at risk of developing a form of cancer can take preventive action, adopting behaviors that have been shown to reduce the risk. Such tests are already available for some fifty hereditary diseases, and more are on the way.

Scientists are also developing ways of treating genetic diseases. Called "gene therapy," the treatments entail transforming viruses into submicroscopic delivery trucks that can transport healthy copies of genes into the cells of individuals who lack them. One such treatment for cystic fibrosis that is now in the development stage involves genetically engineering adenoviruses—the viruses that cause the common cold—so that they each contain a healthy copy of the gene that causes the disease, and then squirting them into a patient's nose. With such developments in the offing, the musings of W. French Anderson, one of the pioneers of human gene therapy, no longer appear farfetched: he envisions the day when a patient will receive an injection of DNA fragments and walk out the door cured.

The pharmaceutical industry has not remained oblivious to the new technology. Many companies are expecting to reap huge profits by developing inexpensive genetic tests. Others are in the business of commercializing gene therapy. Between 1988 and mid-1993, the number of biotechnology companies committed to that goal jumped from four to nearly a dozen. Still other companies are backing researchers who have created transgenic pigs—pigs that have been genetically engineered to possess foreign genes—with the hope that the animals' organs can be used for transplant operations in humans. The federal government, too, has gotten in on the act. In 1990 the National Institutes of Health launched the Human Genome Project, the audacious aim of which is to map (or locate) and sequence (or describe chemically) each of the estimated 50,000 to 100,000 genes in the human genome (the total genetic structure). The project will, among other things, facilitate the search for individual disease genes.

Advances in genetic research have also had an impact on what we eat. For instance, scientists have used recombinant DNA technology to make vitamin C as well as enzymes used in the preparation of food ingredients. In 1994 the first genetically engineered food to be made available to consumers appeared in grocery stores; by altering the gene responsible for rotting, researchers created a tomato that can stay fresher longer. Other scientists have developed a genetically engineered growth hormone that, when injected into dairy cows, results in an increase in the level of the animals' milk production.

Of course, these and similar breakthroughs have raised a host of ethical, legal, moral, and environmental questions. For instance, if an individual is tested to find out if he or she carries a disease gene, will that information be used against him or her by employers or insurance companies? Would some individuals be better off not getting tested at all, given the anguish they might experience if they test positive? Should gene therapy be used to treat only diseases, or should it also be used to enhance an individual's features, such as height or skin color? How safe is it to eat genetically engineered food? Will genetically engineered crops have a deleterious effect on the environment?

While these and other issues will likely continue to be studied and debated for years to come, there appears to be a consensus on one point: there is no turning back. As Francis Collins, the head of the Human Genome Project, has said, "We have already started down this path. Like it or not, we have opened the door and walked through."

PENELOPE BARKER

February 1995

I. GENETICS AND MEDICINE

EDITOR'S INTRODUCTION

Advances in genetic research have had an especially profound impact on the field of medicine. The origins of the revolution in medical genetics have been traced to 1980, when a landmark scientific paper by David Botstein, Mark Skolnick, Raymond White, and Ron Davis was published. The article described a method for finding the approximate location of a gene, and the researchers' key insight was that a gene's location could be deduced by tracing the inheritance of "markers," or distinctive patterns in an individual's DNA, that are inherited along with that gene. "Botstein and Davis's technique was the conceptual engine that drove human genetics from the era of the horse-drawn carriage into the age of the automobile," Robert Cook-Deegan wrote in *The Gene Wars: Science, Politics and the Human Genome.*

The first article in this section, reprinted from the book *Mapping the Next Millenium* by Stephen S. Hall, provides a behind-the-scenes account of how scientists conceived this gene-mapping method. The technique would prove to be crucial to gene hunters who knew the inheritance patterns of a given disease but not the molecular chain of events that led up to it. A case in point was the search for the gene that causes cystic fibrosis (CF), which is the subject of the second article, reprinted from the *New York Times.* The discovery of the CF gene, in 1989, received national attention because it marked the first time that scientists had isolated a disease gene without knowing anything about how it disrupted normal cellular processes. The third article, by Andrew Revkin and reprinted from *Discover,* deals with the discovery of the gene that causes Huntington's disease (HD). This event also generated considerable media attention, because, although the gene was the first to be mapped to a specific chromosome, it took another decade for scientists to isolate it.

In 1990, in conjunction with Department of Energy, the National Institutes of Health initiated the Human Genome Project, a $3 billion project that is expected to take 15 years to complete.

The fourth article, reprinted from *Popular Science,* discusses the origins of the project. This project has facilitated the search for individual disease genes, though its mandate is to map and sequence not just disease genes but the entire human genome, as described in the fifth article, "Taking Stock of the Genome Project," by Leslie Roberts.

Some visionary scientists suspected as early as the early 1950s that if disease had a genetic cause, then curing those diseases would involve repairing the faulty gene or genes that caused them. The historic experiment and subsequent gene therapy trials are discussed in the sixth article by Beverly Merz written in *American Health.* The final article in this section, reprinted from the *New York Times* describes how scientists might successfully apply gene-therapy techniques to the development of treatments for AIDS.

A MOST UNSATISFACTORY ORGANISM[1]

The year was 1978, and the place was a ski resort located in the Wasatch Mountains southeast of Salt Lake City. In April of that year, a University of Utah professor named Mark Skolnick, along with other members of the Utah faculty and their graduate students, retreated to the mountains for their annual spring seminar, and exercising the time-tested sybaritism with which biologists choose their gathering spots, they convened in the lovely setting of Alta. According to the informal agenda, students would describe aspects of their research, and faculty members would comment on it. Out of this modest and free-form scenario would emerge, over the course of a day or two, one of the most powerful technologies in modern biology, a theoretical and yet also cartographic innovation that would forever change the possibility and indeed the likelihood of creating a complete map of human genes. Like many discoveries in science, the Alta breakthrough was built upon a rickety and amusing scaffold of coincidence.

[1]Reprint of chapter 9 from *Mapping the Next Millennium* (1992), by Stephen S. Hall. Copyright © 1991 by Stephen S. Hall. Reprinted with permission of Random House, Inc.

In some ways the coincidences stretch all the way back to Mark Skolnick's childhood in San Mateo, California. It just so happened that Skolnick's father was a social acquaintance of Joshua Lederberg, the biologist who won a Nobel Prize in 1958 for showing that bacteria "conjugate" (a technical term for the exchange of genetic material between biological individuals that in higher organisms goes by other names), so even as a schoolboy Skolnick developed an early fascination for genetics. It just so happened that Lederberg made this discovery about bacteria at about the same time as his colleague Luca Cavalli-Sforza, a prominent geneticist who divided his time between Stanford University and his native Italy; and so it was Lederberg who introduced Skolnick to Cavalli-Sforza when the young man, trained in demographics but enamored of computers, decided to pursue population genetics as a career. It just so happened that Skolnick was living in Italy in the early 1970s, working with Cavalli-Sforza to develop a computer model for population genetics, when he learned by chance about the rich genealogical (and therefore genetic) records of the Mormon population in Utah—a trove of information on family size, births, deaths, and cause of death unrivaled anywhere in the world. It just so happened that the University of Utah turned to Cavalli-Sforza for recommendations when they began looking for someone to manage a program using the Mormon resource; Cavalli-Sforza of course recommended Skolnick, who moved to Salt Lake City in 1974 with the ambition of using those Mormon records to explore the genetic mysteries of carcinogenesis—specifically that of breast cancer.

And finally, it just so happened that in the spring of 1978, when the Utah researchers headed up to Alta for their annual meeting, another of the Utah faculty members had invited along two "yeast men," two high-powered protagonists of the East Coast and West Coast molecular biology establishments, David Botstein from the Massachusetts Institute of Technology and Ronald Davis of Stanford, to sit in on the meetings. In that one conference room would gather all the crucial ingredients for insight: expertise in classical genetics and population genetics, a population to study (the Mormons) with diseases to trace, these two outsiders with their brusque expertise in the powerful new molecular techniques of recombinant DNA, and, no less important, personalities on all sides willing to talk out ideas in rough-draft form and mix it up intellectually. It could have happened, and *would* have happened, somewhere else sooner or later; because of all those coincidences, though, and because of one particular student presentation, it happened at Alta.

During an afternoon session on the second day (as well as anyone recalls), one of Skolnick's graduate students, Kerry Kravitz, presented the results of an extensive investigation into an inherited disease known as hemochromatosis. It is a metabolic disorder in which the body absorbs and retains dangerous levels of iron in key organs and tissues; at least one doctor has marched a patient through a metal detector and set it off, simply to demonstrate how much iron can accumulate. Geneticists knew that hemochromatosis occurred when an individual inherited a defective gene from his or her parents. What they didn't know was whether the gene was *dominant,* meaning a single defective copy inherited from either parent could cause the disease, or *recessive,* meaning defective copies had to be inherited from both parents. And indeed, University of Utah researchers led by George Cartwright and Corwin Edwards had undertaken a classic type of genetic investigation to determine the pattern of inherited disease in an extended, five-generation Mormon family. Then Kerry Kravitz mathematically analyzed that pattern and argued that the data best fit with the hypothesis that the disease gene was an "autosomal recessive." A defective copy had to be inherited from both parents in order for hemochromatosis to develop.

Kravitz and his Utah colleagues helped resolve this longstanding medical controversy with the help of an intriguing fellow traveler to the hemochromatosis gene. As French biologists had reported in 1975, people with hemochromatosis seemed to inherit only one among several possible biochemical "fingerprints" typical of the immune system. It is not critical to understand this system, known as the major histocompatibility complex (or HLA, for human leukocyte antigen); it is only important to remember that these two genetic factors, the hemochromatosis gene and one particular type of HLA fingerprint, seemed to travel together from generation to generation. And on that April day Kravitz described how that HLA protein "marker" had helped the Utah team to identify 20 people with hemochromatosis and approximately 145 others with a single copy of the gene. The "marker," like a red tag or blue ribbon, tipped off the presence of the defective gene that caused the disease. And that gene, Kravitz's analysis showed, was recessive.

As Kravitz explained the statistics used to reach this conclusion, the discussion grew spirited, even loud. The scientists jawed back and forth about the statistical necessity for such complicated analysis. As he followed the argument, David Botstein instinctively found himself thinking about where this information fit in a somewhat bigger picture. His is an especially capacious and eclectic intellect, and so it is not surprising that his thoughts turned to the history of genetic research, to why it was so difficult to study human genetics, and why the Kravitz discovery provided a clue that suddenly, improbably, magnificently promised a way to overcome more than a century of scientific frustration.

"Man," wrote A. H. Sturtevant in 1954, "is one of the most unsatisfactory of all organisms for genetic study."

That is why it has been other organisms—Mendel's peas and Morgan's vinegar flies, Avery's slick-skinned bacteria and molecular biology's viruses and yeast and wriggling nematodes—that have been used to pry out the biochemical secrets of inheritance. For the most universal and obvious of reasons, geneticists cannot put human beings in stoppered flasks, arrange forced matings, create white-eyed mutants, and study the passage of the defective gene through subsequent generations. It would take too long, and geneticists are impatient.

Actually, the secrets culled from lilliputian creatures have a biochemical universality that applies to all organisms, up and down the line. And during the past century, the unstated long-term aim running through genetics has been spatial, if not downright geographic: to plot out inheritance by discovering its organic terrain, surveying it, measuring it, staking out landmarks, and then painstakingly assigning genes to specific regions. Little more than a century after the first fleeting glimpse of that abstract and hidden realm known as heredity, biologists are on the verge of mapping it in its entirety, down to every zip code, every neighborhood, every house, every room.

"It requires indeed some courage to undertake a labor of such far-reaching extent; this appears, however, to be the only right way by which we can finally reach the solution of a question the importance of which cannot be overestimated in connection with the history of the evolution of organic forms." Those words, biology's twentieth-century marching orders, were buried in a paper published the same year that America's Civil War came to an end, words written in German and largely unread (even by Germans) for nearly half a century. They come from Gregor Mendel's classic 1865 paper, "Experiments in Plant-Hybridization."

Mendel, an Austrian monk, was the first to espy biology's Holy Land. During eight patient years experimenting with *Pisum,* the common pea plant, he followed certain physical traits—round or wrinkled seeds, white or dark seed coats, color of pod, length of stem—from generation to generation and discerned that these "constantly differentiating characters," as he called them, seemed influenced by "factors" contributed by the parents. Each parent plant contributed one of these factors in the germ cells, and their combination in offspring occurred as "purely a matter of chance." What he was seeing, in those combinations, was the genetic lottery that leads to round seeds, towering height, blue eyes, mortal diseases like hemochromatosis. Genes were the "factors"; combinations of genes *created* traits. An underappreciated aspect of Mendelian inheritance is the simple but crucial role of *counting*—that is, quantitative thinking based on numbers large enough to be statistically significant. Plotting the percentage of traits from generation to generation allowed Mendel to see the nuts-and-bolts mechanism of heredity. But what exactly were these "factors," and—more important from the geographic point of view—where were they located? What was the domain of the gene?

One of the first hints emerged in 1877, when a German scientist named Walther Flemming visualized chromosomes for the first time in tumor cells, but it wasn't until two stellar intellects congregated at Columbia University shortly after the turn of the century that the picture began to assume some coherence. Thomas Hunt Morgan noticed that mutations in fruit flies—a white-eyed male, to cite his most famous mutant—were inherited in telltale patterns in subsequent generations, depending on the matings (or "crosses") of the parents; in a short 1910 paper in *Science,* Morgan noted that the white-eye trait seemed to be linked to maleness, a seemingly modest conclusion with enormous geographical implications. Maleness in fruit flies, as in humans, is linked to one specific chromosome. Hence, this specific gene was tentatively mapped for the first time to a specific chromosome. One year later, in 1911, Morgan's colleague E. B. Wilson showed that a disease trait in humans, color blindness, occurred only in males, too. It, too, must be associated with the domain that determined sex. Thus was the first human gene, loosely speaking, mapped to the X chromosome (males inherit only one copy of this chromosome while females inherit two).

But it was a sophomore in Morgan's laboratory, the aforementioned A. H. Sturtevant, who discovered the cartographic principle that guides even today's gene mappers. Like Morgan, he realized that certain genetic traits seemed to be inherited in pairs, or linked; unlike Morgan, he went on to examine this linkage statistically and discovered that one could reasonably infer the relative proximity of two genes on a landscape by analyzing patterns of inheritance. That landscape was the chromosome, and the key word in the entire paper was "linear." Sturtevant was saying that the statistical frequency with which two traits appeared together in an organism contained a message about how close the two genes lay next to each other on this linear landscape. Indeed, his 1913 paper explaining this principle contained what can be considered the first genetic map, a short inferential stretch of *Drosophila* chromosome. "As a sophomore in college," David Botstein is fond of saying, "Sturtevant wrote *the* fundamental paper in genetics."

Fundamental, perhaps. All-encompassing? Not by a long shot. It took half a century of superb biology to put some flesh on that thread of insight. In 1944, Oswald Avery's team at Rockefeller University proved that the biochemical substance that transmitted genetic information from generation to generation was deoxyribonucleic acid, or DNA; in 1953, James Watson and Francis Crick showed that DNA assumed the form of a double helix, its genetic information encoded in the sequence of biochemical letters arranged like rungs along the spiraling ladder, its transmission to sex cells (and thus to subsequent generations) facilitated by the unique double helix structure that could chemically separate and unravel down the middle like a zipper; in the 1960s, the genetic code was cracked, showing that each three-letter unit (or codon) of the genetic alphabet spelled out an amino acid, and each gene spelled out a string of amino acids that formed a protein; and between the 1950s and '70s, in a parallel line of research, cell biologists learned to separate and identify all twenty-three pairs of human chromosomes (twenty-four different entities in all, when the X and Y sex chromosomes are counted), lay them out on a page in what is called a karyotype, and stain them to produce unique banding patterns. Each chromosome pair—the twenty-two autosomes shared by men and women, the female XX and the male XY—possessed a distinct size, a distinct pattern of bands, a particular and unique topography.

And so, by the mid-1970s, the heirs of Mendel made landfall on a genetic New World. This New World had a territory, the chromosome. Each of the twenty-four chromosomes, like states in a nation, represented a separate domain, with its own population of genes. Sprinkled throughout the twenty-four chromosomes were 50,000 to 100,000 genes, ranging in size from about 1,500 letters (or "base pairs" of DNA, to use the technical term) to as many as 2 million letters. And those letters of DNA—typically abbreviated as A, C, G, and T for the biochemicals they represent —became a form of measurement, as reliable a yardstick for determining genetic distance as meters or miles on a larger map. Medical geneticists compiled a growing list of inherited disorders that clearly had their origin in genes.

What they *didn't* have up to that point, however, was surveying tools. They didn't have the biological equivalent of theodolite and compass. With almost traumatic swiftness, those tools became available.

If scientific earthquakes could occur in a ski lodge, why not in a delicatessen? In 1972 two biologists broke bread in a Waikiki Beach deli and talked about their work, and by the time Stanley Cohen of Stanford and Herbert Boyer of the University of California-San Francisco finished their sandwiches and their conversation, they had agreed on the outlines of a collaborative experiment that, by 1973, would allow biologists to cut DNA, insert it into bacteria, and copy it. The process became known as cloning. All the elements began to fall into place.

To the molecular cartographer, the tools of the trade include cloning, sequencing, and hybridization, all at the service of several forms of mapping. The first tool available was a kind of biochemical cutting agent known collectively as "restriction enzymes." . . . Each of the enzymes, culled from bacteria, cut DNA with what might be called alphabetical precision; *Eco* RI, perhaps the most famous of these enzymes, cut DNA wherever it found the six-letter sequence GAATTC. You could tell where you were on any given piece of DNA by the locations where *Eco* RI cut it, just as you could tell where you were on a state highway by the location of the county roads that intersected it. Indeed, the literature began to fill up with these so-called "restriction maps." Furthermore, different restriction enzymes cut DNA in different places and with different frequencies, providing something like triangulation and scale; whereas *Eco* RI on average made a cut every four thousand letters or so and was good for high-resolution work, another enzyme, such as *Not* I, cut DNA much less frequently, roughly every 250,000 bases, and thus was better for resolving the large-scale structure of an entire chromosome.

Restriction enzymes did something else, too. They allowed biologists to cut out little pieces of DNA and splice them into the DNA of bacteria, which dutifully reproduced the inserted DNA like a genetic document inserted into a biochemical photocopying machine. Cloning was crucial because biological mappers need copious amounts of the same stretch of DNA to parse out its letter-by-letter sequence and be able to fit it into the larger topography of the chromosome. Finally, in the mid-1970s, Frederick Sanger in England and Walter Gilbert and Allan Maxam of Harvard University independently developed ways to *rapidly* sequence pieces of DNA. With cloning and sequencing, biologists could take any piece of DNA, replicate it, and read the biochemical message of the gene. And since everyone—black and white, man and woman, Bo Jackson and Pee-Wee Herman—possesses the same essential sequence for, say, the insulin gene in his or her chromosomes, each identified gene became a tiny but unvarying piece of the landscape, as distinctive and unique as a genetic mesa, if you will, that appeared in the same place on the same chromosome of every human being. All you had to do was find that place.

We might pause here to propose an extended metaphor that suggests not only the genetic landscape, but the various ways in which mappers attempted to make sense of it. Both the interstate highway system and DNA are double-stranded structures that run through the landscape. Both are *linear* landscapes. An interstate in New York looks about the same as one in California, just as DNA looks essentially the same in chromosome 1 and chromosome 22. So how do you begin to map it and distinguish one part from another?

One way, of course, is to look at the surrounding landscape. Genetic markers can be likened to roadside markers that intersect or lie just off the highway. Restriction enzymes provide another set of markers; they intersect DNA the way exits intersect a toll road, providing one form of geographic precision. Some markers lie near genes, sticking out like mountains or lakes along the roadway, and those form what are called linkage maps. And then the genes themselves form landmarks, which appear large and small along the interstate just like cities and towns; moreover, just as cities reveal their location on the landscape at night with the glow of their lights, genes and other informative stretches of

DNA can be made to glow biochemically in experiments to reveal their location in, say, California rather than New York. That becomes a gene map. By this piecemeal, patient, overlapping kind of mapping, using the powerful surveying tools of modern biology, geneticists can slowly but surely break down the genetic highway system to states, then regions, and finally locate genes between particular markers and particular exits, until all the genes are mapped.

That surveying ability is what molecular biologists brought to the party—a party that, in the sociology of science, they seemed always to be crashing. Flush with the success of their new technological prowess, they would meet scientists from other biological disciplines, tell them how to do their work, and act surprised when they didn't. Yet those new technologies could change the way genetics was done. That is what David Botstein was thinking about—that and the geography of human chromosomes—on the afternoon in Alta when the argument about disease genes and mapping swirled around the room.

"So I listened to the argument," Botstein recalled, "and finally I got in the middle of it." For anyone who knows Botstein, this is not an uncharacteristic set of conversational coordinates. Well spoken, good-humored, with sardonic and refreshing opinions on almost everything, he had early on established a reputation as a leading light in molecular biology, although his first love was music and he had intended to pursue a career in choral music or as a conductor. (He comes from a family of polymaths; both parents are physicians, and his brother Leon, currently professor of history and president of Bard College in New York, was recently named conductor of the American Symphony Orchestra.)

Botstein heard Kerry Kravitz's comments with the ears of a molecular biologist; but he also heard them with the ears of someone steeped in the history of genetics. The early work of Sturtevant and Thomas Hunt Morgan, who had begun to infer the spatial arrangement of genes in fruit flies by studying patterns of inheritance, was "mother's milk" to geneticists, as he put it later. The discussion at Alta became too complex to report here, but it would be fair to say that Botstein immediately realized that the paired appearance of hemochromatosis and its HLA "flags" could—*could*—serve as a new way of establishing recognizable landmarks on all the human chromosomes, landmarks almost as regular as mileage markers along a highway. And the key to finding those landmarks was something called polymorphisms.

"Polymorphism" is one of biology's less friendly words of jargon. But the concept is quite simple. It refers to a patch of genetic landscape—the fringe of a gene, a random stretch of letters, part of the gene itself—that contains generally insignificant but discernible variations. A classic example is blood type. Everyone has the same human genes for hemoglobin, but small variations result in one of four standard blood types: A, B, AB, and O. Those are polymorphisms—slightly different forms of the same thing. If people with O-type blood happened to be the only people to develop hemochromatosis (which is not the case), then geneticists would say the genes for the disease and the O polymorphism were "linked," implying that they lay in the same genetic neighborhood.

Now Botstein knew that inheritance wasn't quite as neat and clean as Mendel's pea plants implied. Just to confuse matters more, the landscape of heredity rearranges itself a bit. Chromosomes swap and exchange segments during the formation of sex cells. Just prior to migrating into sperm or egg sex cells, while they are lying side by side, chromosome strands often cross over each other, the way lovers sometimes cross legs while asleep at night; unlike the legs of lovers, those chromosomal limbs switch places, merging onto the adjoining strands. This obviously confuses both landscape and linkage. The farther apart two genes are, the likelier they are to become separated at some point during this process, known as "recombination."

Botstein didn't need to explain all this to the Utah group. But he talked about how mathematical analysis could penetrate that confusion. "And then I said—and this I'm sure I said exactly this way—I said, 'If you had things as polymorphic as HLA all over the genome, then you could map anything.' And just as I said that, I was looking at Ron Davis. Ron Davis is looking at me. And as the words come out of my mouth, we both realize that in fact we *have* a way of doing that, because we're working on it!"

Molecular biology, in short, offered an end run to this problem. Polymorphisms don't occur only in genes. Working with yeast, Botstein and Davis had seen polymorphisms pop up on the fringe of yeast genes, and the variations revealed themselves when you cut the yeast DNA with restriction enzymes. Some of the fragments were long, as if they had tails; others were short. If a long-tailed marker came from father, for example, then the defective gene linked to it came from father, too, and patterns of inheritance suddenly, magically became clear. The idea would become known as restriction fragment length polymorphism, or RFLP, denoted by the neologism "riflip." These "riflip" tails, of varying lengths attached to genes, flagged genes in a way that allowed them, and genetic diseases, to be traced through generations.

All during the day, they talked out the idea. Davis, soft-spoken and highly respected, agreed that the technique should work. As Skolnick remembers it, "It took a while before the ideas gelled." But in Botstein's memory, the genetic seas parted then and there. "By dinner that night," as he puts it, "we had laid out the whole business of how you would make a map—Skolnick and Davis and myself, with kibitzing from a few others. So this idea—for *us,* in any case—was generated at that moment. And basically the development of the next five to ten years was clear."

The future of genetic cartography may have been clear from that moment, but not until two years later—a lapse of time almost geologic given the pace of discovery during that period of molecular biology—did Botstein, Davis, and Skolnick publish the landmark paper outlining this strategy. It appeared in the May 1980 issue of the *American Journal of Human Genetics,* and there was a fourth author, Raymond L. White. A researcher at the University of Massachusetts Medical School at the time, White was not initially impressed with the idea; as he recalled later, "My initial response was one of intellectual irritation at the grandiose and presumptuous scale" of the strategy. Yet he volunteered to take on the task of proving that RFLPs could indeed serve as useful markers. The others waited until White and colleague Arlene Wyman found their first mappable marker, which they did by late 1979. White then moved to Utah to begin constructing an RFLP map with Skolnick, with the Mormon resource providing an unusually rich lode of potential landmarks.

In the uncharted terrain of chromosomes, "riflip" markers revolutionized the mapping of that landscape. Botstein sent a preprint of the RFLP paper to several prominent human geneticists, including Victor A. McKusick of Johns Hopkins University School of Medicine. McKusick, author of *Mendelian Inheritance in Man* and unofficial Mercator of the human genetics community since the 1950s, realized that RFLPs gave to gene mappers what the longitudinal clock gave to old geographers: precise meridians marking the landscape. The technique "opened up tremendous possibilities. It was a very exciting, landmark, watershed paper." In 1980, when the paper was published, McKusick's genetic gazetteer consisted of "at most two dozen markers. Now," he added, speaking in 1991, "we have twenty-four hundred markers. That gives an idea of what sort of explosion we've had, and it's been not only a quantitative explosion but a qualitative one."

That was the productive aspect of the RFLP paper. Unfortunately, the opening of a new territory, not for the first time in the history of cartography, led to territorial skirmishes, commercial competition, and political disputes, the effects of which are still evident within the biological community. Or, as Botstein puts it, "The history thereafter is a little bit less . . . attractive."

With the powerful surveying tool of polymorphic markers, the question arose: should biologists try to find specific, high-profile genes related to human disease, or should they take a more systematic approach, create markers throughout the genome (the total of all human genetic terrain), and in effect lay down a preliminary grid of genetic coordinates? In 1983 that question seemed to be decisively answered when James Gusella and his coresearchers at MIT located an RFLP marker for Huntington's disease. This devastating neurological disorder unfolds in a particularly tragic manner, for it reveals itself only in the fourth, fifth, or sixth decade of life, after the childbearing years when the mortal dominant gene may already have been passed on to children.

The marker wasn't the gene, mind you. It simply indicated the right genetic neighborhood. Nonetheless, the Huntington's discovery set off a gold rush for disease genes. "People reacted the same way they do in Las Vegas when they see a slot machine pay out lots of money," says David Botstein, who had been arguing for years that a modest RFLP map of all human chromosomes could be made for about $5 million. "What they do is they *cluster* around the machine and put more money in—a phenomenon that the people who *own* the casinos have exploited to their eternal, endless profit. That's what happened to human geneticists. They started to play the slot machine."

A few hit the jackpot. In the 1980s, genes linked to such prominent disorders as muscular dystrophy, cystic fibrosis, and chronic granulomatous disease were discovered and mapped. But at the beginning of 1991, the Huntington's gene still eludes researchers, who have been wandering in the wilderness at the very tip of chromosome 4, where the gene is believed to lie. [Editor's note: The Huntington's gene was located in 1993.] Between 1983 and 1990, the National Institutes of Health (NIH) spent $15.2 million on the search for this one gene; for $15 million, Botstein has argued, a reasonably detailed RFLP map of *all* the chromosomes, with landmarks closer together than researchers currently are to the Huntington's gene, could be made. With each passing year, the argument grows more persuasive, the approach certainly more systematic and cost-effective.

There was another, more political barrier to speedy progress. The push to map human genes emerged not from the human-genetics community, but from those pushy molecular biologists. But the human-genetics community—through its funding section at the NIH—was slow to grasp the cartographic power of markers, according to several key biologists. "What they were particularly skeptical about—and this skepticism remains—is the value of a *map* as opposed to just individual markers," Botstein says. And that skepticism accounts for the rather unusual fact that the two earliest mapping initiatives were supported not by the NIH's customarily astute funding mechanisms, but by private enterprises: a philanthropic organization in one case, a small biotechnology company in the other. And it was the French who most successfully pushed for a cooperative international mapping effort. In 1984, Nobel biologist Jean Dausset formed the Centre d'Etude de Polymorphisme Humain (CEPH) as a central clearinghouse for genetic landmarks; researchers would deposit markers and cell lines from families, such as the Mormon families in Utah, and any researcher could then request DNA from the CEPH collection in the interest of creating a linkage map of human genes.

When it became apparent that the NIH balked at supporting a mapping project—it looked like "a fishing expedition" at the time, McKusick admits—Botstein and others "shopped the idea" to outside sources of funding. The Howard Hughes Medical Institute ultimately agreed to fund a group at the University of Utah—an obvious place because it had the Mormon resource at its fingertips, the acknowledged pioneer of RFLP mapping in Ray White, and Mark Skolnick's expertise in population genetics (White and Skolnick, however, would later have a falling-out). White's group promptly set out to map individual chromosomes, beginning with the short arm of chromosome 11, while taking productive detours to study interesting disease genes.

Meanwhile, Botstein and others pushed an effort in the private sector, and that is how the first overall map of markers and genes on human chromosomes began to take shape on a corkboard office wall in the suburbs of Boston. It was there, at a small biotechnology company called Collaborative Research, that Helen Donis-Keller and a team that would ultimately number thirty-three biologists began to attach little flags of blue, yellow, and orange paper to long vertical strands of blue and red yarn pinned to the wall; each strand of yarn represented one of the twenty-four human chromosomes, male and female, and the tags represented genes and RFLP markers. Starting with 12 useful markers in 1981, the Collaborative Research team gathered 403 markers, including 393 "riflips," and began plotting them. Eric Lander of MIT invented a computer algorithm called MAPMAKER that translated statistical linkage into genetic geography.

In October of 1987, the Collaborative team announced it had completed the first map of the human genome. The map appeared on the cover of the journal *Cell,* the undisputed Bible of molecular biology, and was greeted by a tremendous reaction ranging from praise to censure. Praise because it was a first draft of human genetics, however crude and full of gaps (which it was). Censure because some claimed it was too incomplete and borrowed heavily on the work of others—specifically markers donated to CEPH by the Utah group. Another undercurrent of discontent stemmed from the fact that because Collaborative Research wanted to patent some of its markers for genetic-testing kits, the company stood to benefit commercially from researchers with whom it shared markers. The dispute surfaced in an unusually public display of enmity between research groups which, amid the rhetoric and ill will, raises some interesting philosophical questions about what constitutes a map, when is the proper time to make one, and by what authority its utility should be judged. Recalling the ardor with which Strabo attacked the irregular and incomplete meridians of Eratosthenes' third century B.C. map of the world, it is by no means a new debate, only a new territory.

Ray White described the Collaborative map as "premature and presumptuous" in a newspaper interview, and told Leslie Roberts of *Science* magazine, "It is a very useful collection of markers, but it is not what we believe should be properly called a map." In *Science,* Helen Donis-Keller replied, "A map is a map.

Our map has holes, we make no bones about it. This is *a* genetic map of the genome. It is not Ray White's ideal, but so what? This is the beginning for us. How can one person set the standard for the rest of the world on what constitutes 'the map'?"

Like all maps, the Collaborative map invited further exploration, higher resolution, and more dedicated effort. But the dispute did leave some hard feelings. Before she moved on to Washington University, a white sweatshirt hung in Donis-Keller's Massachusetts office, a gift from her research team. In black gothic letters, it read simply PREMATURE AND PRESUMPTUOUS.

Even while that first crude map was being assembled, David Botstein and another dozen or so biologists gathered at the University of California-Santa Cruz in May of 1985, invited by the campus's chancellor, Robert Sinsheimer, to discuss the most ambitious genetic mapping project of all: to map and sequence the entire human genome.

Botstein was there, and Donis-Keller, and Leroy Hood of Caltech, and Walter Gilbert of Harvard—heavy hitters all. Unlike many scientific meetings, this one began with a consensus. "Most people thought the idea was crazy," Botstein recalls. But as the reports of technical progress came out during the informal two-day meeting, the mood shifted. As Sinsheimer later summed it up, "There was perhaps a consensus that it could be done, but not that it *should* be done."

From that moment on, after the leading lights of the field convinced themselves of its feasibility, the project gained the kind of intellectual bona fides that carry tremendous weight in the scientific community. This is not the occasion to describe the many meetings and workshops, technical arguments and battles over turf, that ensued; the story is only now beginning to unfold. But by October 1, 1989, barely a decade after the original RFLP paper appeared in the literature, the $3 billion Human Genome Project was officially up and running.

The genome project will unfold in a series of maps of ever-increasing resolution. The first crude RFLP map, its four hundred or so markers spread over twenty-four chromosomes, represents little more than a hand-drawn sketch of the territory, divided by forlorn and infrequent intervals similar to lines of latitude and longitude. A physical map, currently being assembled, is also a map of broad brush, but it might be likened to breaking up the chromosomal territory—the interstate genetic highway, if you will—into smaller, self-contained segments, such as the part of Interstate 80 that runs through Iowa. The genetic map, more particular still, provides the exact location of genes—pinpointing the exact location of Des Moines and Davenport, for example, along that particular stretch of I-80. The color map illustrating this chapter shows a technique developed by workers at Yale University by which biologists can assign a small physical part of the map, a "segment" of the genetic highway, to its exact location on its particular chromosome. The final map, the map providing ultimate resolution, is the equivalent of a street map of Des Moines or Davenport. This is the sequence—the fine-scale topography of the genes, as spelled out in the sequence of biochemical letters in the genetic code. As information increases, diagnostic tests will be developed, genes will be studied, disease mechanisms explored, and—it is hoped, but by no means guaranteed—cures may be discovered.

Mark Skolnick, it will be remembered, moved to Utah with the dream of finding the genetic causes underlying breast cancer. In 1974, before the molecular surveying tools were in place, that may have seemed little more than a pipe dream. Finally, in 1990, Skolnick and his team converted the dream into partial reality; they published a paper strengthening the case for a genetic predisposition to breast cancer in women, which raised hopes that such inherited tendencies might be identified ahead of time. It was merely the latest crest in what he calls a "tidal wave" of change in the way human heredity is being charted and understood. Over the next decade or two, those changes will affect the way we think about having children and the way pregnancies proceed, the way we make decisions about our long-term health and the way we apply for health insurance (and why we may be turned down).

In that future, as Skolnick sees it, "clinicians doing diagnoses will request sequence testing the way they do allergy tests today. On a chip-sized blot, one centimeter by one centimeter, just like an integrated circuit, you'll be able to test for ten thousand genetic sequences. We'll know so much sequence information by then, so much about normal and so-called abnormal sequences, that we'll be able to identify tendencies to disease before they occur. And then all of medicine will be completely different. You'll be born, you will have your sequences tested, and if there are sequences associated with disease, we'll try to counter those before there are symptoms or cell degradation."

And how far away is this?

"Not that far," says the forty-five-year-old Skolnick. "I think I'll see it in my lifetime."

TRACKING DOWN DEFECTIVE GENES[2]

Researchers have developed a battery of new techniques to track down and identify defective genes that give rise to thousands of hereditary diseases.

In a stunning first example of putting the techniques to work, a team of Canadian and American scientists reported last week how they "walked" and "jumped" up and down a human chromosome to zero in on the gene responsible for cystic fibrosis. The report is in the current issue of the journal Science.

Other researchers say they are applying the techniques to Huntington's disease, neurofibromatosis, or Elephant Man's disease, and a host of other conditions.

The techniques pave the way for a new era of what researchers call "reverse genetics," said Dr. Victor A. McKusick, a professor of medical genetics at Johns Hopkins School of Medicine who is an authority on gene-mapping techniques.

Each person has 50,000 to 100,000 genes arranged along 23 pairs of chromosomes. Each gene generally produces one of the many proteins that direct life processes.

In conventional genetic research, scientists try to identify the gene that causes a particular ailment. But they can do that only if they know what protein abnormalities are involved. Once they identify the protein, identifying the gene that produces it is relatively straightforward.

For example, sickle cell anemia is characterized by abnormally shaped red blood cells. Because they knew what cells were abnormal in the disease and what protein was likely to be the culprit, researchers could deduce that the problem involved the gene that produces hemoglobin, the oxygen-carrying protein in red blood

[2]Reprint of article by Sandra Blakeslee from the *The New York Times* p C3 S 12 '89.

cells. When they looked at the gene in people with the disease, they saw an abnormality.

Such "forward genetics" has been successfully applied to only a small fraction of the estimated 4,000 diseases caused by single defective genes, said Dr. Francis Collins, an associate professor of internal medicine and human genetics at the Howard Hughes Medical Institute and University of Michigan.

Reversing the Search

But in most diseases caused by defects in single genes, these techniques are not easy to apply, usually because scientists do not know what proteins are involved or where they are produced in the body. So they use reverse genetics to "go first for the gene, then work back to the disease," Dr. McKusick said.

The techniques have thus far successfully located the defective genes responsible for several inherited diseases, including Duchenne's muscular dystrophy.

In these diseases, however, researchers already had important clues to the location of the genes involved. For example, scientists knew that the muscular dystrophy gene was carried on the X-chromosome, which had gross abnormalities in people with the disease.

No such clues existed for cystic fibrosis, a disorder of the breathing system, pancreas and sweat glands and the most common genetic disease among Caucasians. About 30,000 people in the United States have cystic fibrosis, and 1,000 to 1,200 new cases are diagnosed each year.

Dr. Collins and his Canadian collaborators, Dr. Lap-Chee Tsui and Dr. Jack Riordan at the Hospital for Sick Children in Toronto, said searching for the cystic fibrosis gene was like looking for a broken faucet in a house somewhere in the United States. The house is the gene. The broken faucet is the defect in the gene.

"When we began our search seven years ago," Dr. Collins said, "we had no idea what state to look in. The house could have been anywhere between New York and San Francisco."

The first technique they used was genetic linkage analysis. Researchers studied the chromosomes of hundreds of families affected by cystic fibrosis, searching for patterns. They found that everyone who had the disease carried a specific pattern on chromosome seven. Family members free of the disease did not show this pattern. [Editor's note: The development of this technique is the subject of the first article in this section.]

This was like determining the house is located between New York and Boston, Dr. Collins said.

A second technique, called pulse field electrophoresis, helped determine more precisely where on the chromosome the defect was and how many base pairs, the chemical constituents of DNA, were involved.

By the summer of 1987, Dr. Collins said, "we knew we had to search 1.7 million base pairs to find our house." He continued: "The search had narrowed from New Haven to Hartford. It was a daunting task. We needed new technologies to do it."

Random Segments Mapped

The researchers used a new technique called chromosome jumping. "We were not interested in walking down every street looking for our house," Dr. Collins said. "Instead, we jumped from one end of each town to the other end, asking—yes or no—is the house down there?" They made the jumps by comparing blocks of about 100,000 base pairs to DNA isolated from people who carried the cystic fibrosis gene, looking for the closest possible match.

Dr. Tsui employed another technique called saturation mapping to help identify some of the more interesting villages. He took 250 random segments of DNA from chromosome 7 and compared them with DNA from cystic fibrosis carriers. Two segments consistently tended to match.

At this point the scientists joined forces. "Our search parties covered the territory a lot faster," Dr. Collins said.

Last January, they found a stretch of 300,000 base pairs that matched almost perfectly with DNA from cystic fibrosis carriers. "We knew the cystic fibrosis gene had to be in there somewhere," he said.

This segment contained enough DNA to account for several genes. "It was time to start walking through the streets looking for the house," Dr. Collins said.

Several techniques were used. One involved snipping the DNA into small pieces, tagging it and seeing if it made RNA—the fragment copied from DNA that organizes amino acids into proteins inside a cell. Every cell contains 10,000 to 50,000 RNA molecules, Dr. Riordan said, each governed by a different gene.

The problem was that researchers did not know where to cut the DNA. Since 85 percent of all DNA is inactive, Dr. Collins said, most fragments contained non-functioning DNA.

"We almost missed the gene when we were in it," he said, "because we cut in the wrong place."

Important Genes Found

Nevertheless, a process of elimination led to one length of DNA. It was assessed with another technique called a zoo blot: scientists compared the human DNA with samples of DNA taken from other animals. The human DNA is likely to contain an important gene if it matches up with DNA segments taken from many species, the theory being that important genes are conserved throughout the animal kingdom.

"We had to believe it was a gene," he said, "but we were still not sure. If it was a gene it should make a protein relevant to cystic fibrosis. But what protein?"

To find out, Dr. Riordan of the Toronto team constructed a copy-DNA library, a collection of all the RNA molecules found in a single cell. Dr. Riordan and his colleagues made their library out of the RNA found in human sweat gland cells because faulty sweat glands are a hallmark of cystic fibrosis. If researchers had the correct house, one of the RNA molecules in the library should match the street address they had found.

The task was to compare about 10 million pieces of copy DNA with the candidate gene found in the village. After eliminating countless false matches, he said, they hit the jackpot: a gene that made a membrane protein in sweat glands.

Now the researchers had to discover how the gene failed in people with cystic fibrosis.

They compared the DNA sequence of the gene from people without cystic fibrosis with the same DNA taken from cystic fibrosis patients. Using another technique called polymerase chain reaction to amplify their search, the researchers hit the prize. A third of the way through the house, researchers found the broken faucet: three base pairs of DNA were missing in those who had cystic fibrosis.

HUNTING DOWN HUNTINGTON'S[3]

Joseph Hartman and his wife, Marilyn, smile stiffly as they sit in the Ben Franklin room of the University of Pennsylvania's student union. He is gaunt and pale, with a long face and cropped, graying hair. She has the delicate frame of a songbird and is prim and neat in glasses and a green and white floral-print dress. They might be posing for a picture—a modern version of *American Gothic*—except that Joseph Hartman's body is in constant motion.

Two dozen physicians and biologists politely but intently inspect Hartman as his case is presented by his neurologist. Hartman (not his real name) first sought medical help 13 years ago, complaining of irritability, lack of sleep, the jitters. He soon became unable to work and began to lose his short-term memory.

Only after many years did the chorea begin. The word comes from the Greek for "dance," and that is exactly what Hartman is doing now. His hands dart and float in the air as if manipulated by an inebriated puppeteer. His head has a looseness that recalls those toy puppies whose heads once bobbed in the rear windows of American sedans. He watches his own limbs as if from a distance, occasionally reasserting control long enough to force a hand to grasp at the fabric of his pant leg or scratch the nape of his neck. Inevitably, his body rebels and resumes its dance. It is a dance of death.

This is Huntington's disease: a devastating inherited condition that often waits until midlife to strike. A rare genetic flaw, present in one of every 10,000 people, selectively destroys two small regions of the brain—the putamen and the caudate nucleus—that help control movement. Eventually the muscles cannot be controlled at all; many Huntington's patients die because they choke on food they can no longer swallow.

The presentation of Hartman's case is the opening act in a two-day workshop aimed at generating research on how the gene kills brain cells. Hartman and his wife respond to questions. "Do

[3]Reprint of article by Andrew Revkin from *Discover* v 14 pp98–106+ D '93.

you like sweets?" asks the woman seated next to Hartman's wife. The woman could not present a more dramatic contrast to Joseph Hartman. Her blue eyes are clear and probing; she holds her shoulders square and steady. Indeed, Nancy Wexler, a psychologist and president of the Hereditary Disease Foundation, which organized the session, is as elegant and sharp as the tailoring of her cream suit.

After a pause, Hartman replies, "I like fruit." He speaks in a shallow voice that seems only marginally under his control. "I've practically given up candy. My grandchildren say my loss of candy is not a good thing." The joke elicits polite chuckles from the scientists.

Wexler comments to the gathering that Huntington's victims often crave sweets and rich foods, perhaps to provide the extra energy it takes to keep the body in perpetual motion. "Sometimes they'll get up in the middle of the night and fix themselves enormous dinners," she notes.

Wexler's familiarity with Huntington's is not merely academic. Twenty-five years ago, when she was 23, Wexler learned that her mother, Leonore, had Huntington's disease. That means there is a fifty-fifty chance that she has inherited the genetic mutation and will eventually die just as her mother finally did—just as the man in front of her is slowly dying now.

The specter of the disease has dogged Wexler every day since 1968—"Any time I trip on a curb or drop something, I think, 'Is this it?'"—but she, in turn, has hounded the disease. For more than two decades she has devoted all her energy to finding the destructive genetic misprint and a way to cure it.

In March of this year, a consortium of 58 researchers—including Wexler—announced that they had accomplished the first goal. The "gene hunters," as the group was informally called, had discovered Huntington's lethal mutation: an extended, stuttery repetition of a single DNA instruction. Now many labs are racing to create cell cultures and gene-altered mice that may lead to treatments or, eventually, a way to repair the defect through gene therapy.

Wexler has hardly paused to celebrate. The Philadelphia workshop is the first of a new wave of "post gene" workshops, one in a series of dozens of similar sessions she has organized over the years. When the presentation of Joseph Hartman's case draws to a close, Mrs. Hartman grabs some muffins and a glass of orange juice for her husband. Wexler glances at the food and pats Joseph

Hartman's stomach. "Keep eating!" she says, and hugs him tightly, as if he were a long-lost relative whom she might not see again. In a way, he is.

It is almost impossible to talk about Huntington's disease without talking about Nancy Wexler. In the fight against the disease, she has functioned not only as scientist but as catalyst, cheerleader, even den mother. Her quest has taken her from lavish fund-raising dinners to congressional hearings, from genetics laboratories across North America and Europe to squalid shantytowns on the shores of a Venezuelan lake. Her personal struggles—with her mother's illness, with the decision about whether to be tested for the gene—have only added to her credibility. And through it all, she has remained an eloquent advocate for victims of the disease. In September, she received the prestigious Albert Lasker Public Service Award in recognition of her work on their behalf.

Yet when asked, Wexler credits her father with sparking her crusade. "There's something very fundamental about how a family or any human being faces a critical problem," she says. "The choices in my family have very much been determined by my father's reactions to the crisis—how he dealt with it."

Just a few minutes apart, on a hot August day in 1968, Nancy and her sister, Alice, arrived in Los Angeles, answering their father's invitation to celebrate his sixtieth birthday. Nancy thought something was up; theirs was not a family that was big on birthday parties. Milton Wexler was a tall, urbane man who had succeeded in three careers—as a lawyer, a commander in the Navy, and a psychoanalyst. He has since added another, collaborating with friend Blake Edwards on several successful screenplays. Alice, then 26, was well on her way to a doctorate in history at Indiana University; today she is writing a book about Huntington's disease. Nancy had recently graduated from Radcliffe and was about to begin graduate studies in psychology at the University of Michigan.

Their father did not tell them anything until he had driven them back to his apartment. There he broke the news that their mother, from whom he had been divorced for several years, was doomed. (In retrospect, he says, many of the behavior patterns that led to the split—depression, irritability, and difficulty relating to others—read like a textbook on the early symptoms of the disease.)

Nancy says she doesn't really remember much else about that

day, except that she knew her mother was dying and that she and her sister decided never to have children. Milton tried to tell them what was known about the disease, but there wasn't much to tell. Huntington's afflicts 30,000 people in the United States and places another 150,000 at risk. The gene that causes the disease is dominant, meaning that if a child inherits just one copy of it, he or she will get the disease. As Milton spoke, Nancy's mind rushed back to the early deaths of Leonore's three brothers. "The brother closest to her, Seymour, was a fantastic clarinetist," she says. "When he died, Mother was very upset and went back east. I asked what was going on, and I was told that all of her brothers had died of a hereditary disease, but that we were not at risk."

Once the shock had passed, Nancy, Alice, and Milton rallied around Leonore. But they didn't stop with that. Milton got in touch with the widow of the disease's most famous victim, folk singer Woody Guthrie. Marjorie Guthrie was organizing victims' families into the first organization to call for research, the Committee to Combat Huntington's Chorea, and Milton decided to form a California chapter of the group. Guthrie's organization, however, focused on building state chapters and lobbying; Wexler wanted to fund scientific research directly. "My father felt very strongly—because it was such a devastating disease—that even though care could certainly be improved, the crucial thing was to find a treatment that took away the symptoms and cured the disease," says Nancy. "That no matter how much you padded a bed with lamb's wool so someone wouldn't bash their legs and get black and blue on the railings, it was much better to get them out of bed. And that's been the guiding philosophy."

Eventually the differences forced a split—in 1974 Milton Wexler created the Hereditary Disease Foundation. Even before then, he began pursuing his scientific agenda by setting up a series of freewheeling discussions on Huntington's; he lured the best and brightest young minds to these workshops by offering free travel and $1,000 honoraria.

He was frustrated by the first meetings, at which young scientists gave old-fashioned presentations bogged down by slides and charts. So he refashioned them. From then on, sessions were held in the round, with no set agenda. "What we ask for are people's imagination, energy, and affection for a weekend," Nancy Wexler explains. "If, however, anybody feels inspired to actually tackle some of these problems, we're quite interested in knowing that."

Allan Tobin, a young neuroscientist from Harvard who knew

Wexler from her undergraduate days, says the early workshops were like a game of "the blind man and the elephant," where no speculation was too wild. In addition, he fondly recalls how Milton Wexler would sometimes use his contacts in the entertainment community—many of whom also happened to be his patients—to arrange a Hollywood-style party as the centerpiece of meetings held in Los Angeles. There the researchers hobnobbed with the likes of Jennifer Jones, Candice Bergen, and Cary Grant.

Eventually Tobin moved to UCLA and became scientific director of the Hereditary Disease Foundation. He also became the moderator for the Huntington's workshops, refining a style that one of his colleagues calls "a cross between Socrates and Geraldo Rivera." Over the years he and Nancy Wexler have developed the habit of sitting side by side while they watch the assembled scientists play intellectual volleyball. "That's so I can reach out and kick him if he intrudes too much," she jokes. "And vice versa."

Wexler found her true mission, however, not at one of her own workshops but at an outside meeting. It was 1972, and she was attending a symposium marking the 100th anniversary of Long Island physician George Huntington's landmark paper describing the disease that would come to bear his name. At the meeting Ramón Avila Girón, a Venezuelan psychiatrist, talked about a large population of Huntington's cases clustered along the shores of Lake Maracaibo, a 130-mile-long brackish lake best known for the vast oil reserves beneath its muddy bottom. Then he turned down the lights and showed a grainy black-and-white film, replete with a tinny sound track of patriotic music.

One after the other, people walking down the streets or sitting in cafés danced to the tune of the Huntington's gene. Inbreeding, isolation, and a tendency toward large families—one woman had 18 children; one man, 34—had produced an extraordinary, possibly unique, concentration of the mutant gene. "It was a total shock," Wexler says. "Here were all these Huntington's cases, practically in every household, not shut away in nursing homes like they are here, not being stared at, but accepted as part of a community."

Though Wexler felt drawn to these people, it would be several more years before she'd have the chance to meet any of them in person. In 1974, after completing her doctoral thesis on the psychology of people at risk for Huntington's disease, she moved to New York City to teach. Less than two years later she was named executive director of a new congressional Huntington's Disease Commission to set priorities for federal funding to fight the disease, and she moved to Washington, D.C. At this point Milton Wexler realized that it was time for him to step back a bit from the project he had started: the science was getting more complicated, and Nancy and Allan Tobin had the workshops under control. "From there on," says Milton, "the story is Nancy's."

One of the first things she did was assemble a Venezuela Working Group so that she could follow her instincts and see that the lakefront population was studied. In 1977 her commission gave its recommendations to Congress, and federal funding began to flow to the gene hunt.

Her triumphs were tempered by her mother's death, on Mother's Day of 1978. Wexler sadly recalls visiting her mother in nursing homes as the disease took its toll. "She was extremely frail. All of us were always afraid when she took a step that she would go careening forward onto the concrete. It wasn't a soft environment; everything was unfriendly—concrete floors, hard walls, the chairs weren't soft, the bed wasn't soft."

Spurred by the loss of her mother, Wexler made preliminary trips to Venezuela in July 1979 and April 1980. In March 1981 she made the first of what would become annual expeditions to collect blood and chart this sprawling Huntington's family tree. The lineage is now the largest ever documented, numbering more than 13,000 individuals. Day after day, assailed by insects and heat, she and about ten other researchers explored the barrios around Maracaibo and traveled to outlying villages. They worked in local dispensaries or government-built clinics. "It was a sauna," Wexler says, describing one of these facilities. "We had to scream over people's heads. The room was packed because this was such a novelty." The Venezuelans were not the only ones to find the experience both odd and compelling. "Here was this setting that couldn't have been more different from anything I'd seen in my life, and yet here was this totally familiar disease," Wexler adds. "I was exhilarated and frightened. I felt connected and alienated."

She and her colleagues wanted to take blood samples to use in studies of how Huntington's does its damage to the body, but because few of the people had ever had blood drawn, they were scared. "It was hard to describe why we wanted the blood of healthy as well as sick people," Wexler says. She explained over and over again that the large family tree around Maracaibo could provide special clues to a disease that was hurting people around the world. "I told them my mother had this," she says, "and that I was at risk. I told them that very far back we were family. They felt that bond." She returned to the United States with blood samples, a crude pedigree, and high hopes.

Those hopes were bolstered by an ongoing revolution in molecular biology. For several years researchers had been refining new tools called restriction enzymes; these enzymes, they believed, which could be used to snip DNA into manageable pieces, would someday let them pick out specific disease-causing genes from the huge, bewildering tangle of DNA that makes up a person's entire genetic endowment. Each restriction enzyme recognizes its own specific sequence of four to eight nucleotide bases—the building blocks of genes. Wherever it spots that sequence, it makes a cut in the DNA. Over a long stretch of DNA, a restriction enzyme will excise snippets of varying sizes, depending on the number of nucleotides sitting between its cutting sites. Once researchers have the bits of DNA, they can copy them and determine the order of their bases—that is, they can sequence the gene or genes contained in that DNA.

Although more than 99 percent of human DNA is exactly alike from one individual to the next, at certain spots along the chromosomes short stretches of DNA display distinctive variations, called polymorphisms, and these differences are passed within a family from one generation to the next. The variations may mean that a cutting site that exists on one person's chromosome is missing from the same spot on another person's chromosome, or that extra nucleotide bases are added between two cutting sites on one person's chromosome but not on another's. In either case, when restriction enzymes go to work on the same chromosome from two different people, the resultant fragments may vary in length or weight, and so they can be used to distinguish one person's DNA from another's.

It was quite possible, the reasoning went, that some of these easy-to-spot polymorphisms sat very close to a disease-causing gene on a chromosome. If so, then the polymorphism and the gene would be likely to stay together, even when the chromosomes get shuffled around, as they do when eggs and sperm are produced. In other words, if you could find a particular polymorphism that consistently traveled with a particular disease, you could use that polymorphism as a marker to tell you that the disease gene was located somewhere nearby. [Editor's note: A discussion of how scientists conceived this technique is provided in the first article in this section.] And you'd know that, in any given family, anyone who had the polymorphism would be at risk for the disease.

When Wexler first encountered the idea of using polymorphisms as markers, it was just that—an idea. It was October 1979, and she was hosting yet another workshop. She listened intently as key theorists in the field explained their vision for the future of gene hunting. "All the people there were true believers," she says. "Once you accepted the premise that you really could find markers, then it was just a matter of time to find the gene. It might take you a million years, but it wasn't a complete wild-goose chase where if you miss then you end up with nothing. The whole idea looked beautiful."

One of that workshop's leaders was David Housman from MIT, a friend of Tobin's. After the meeting, he returned to Cambridge and persuaded one of his graduate students, James Gusella, to focus on finding Huntington's markers. Gusella, who soon graduated and moved to Massachusetts General Hospital, began identifying and collecting DNA probes—bits of DNA that are made up of the nucleotide bases complementary to those found around a particular polymorphism. A good probe would latch onto only the fragment of DNA containing that polymorphism. Ultimately, of course, the probe Gusella wanted to find was one that latched onto a polymorphism inherited along with a Huntington's gene.

Collecting the probes was slow work. By 1982—when researchers had discovered only a few dozen polymorphisms—Gusella had his first batch of 13 probes ready for testing, with more waiting in the lab. He was not too hopeful; he noted that it could take 300 probes to more or less cover the entire human genome and find a marker within a reasonable distance of the Huntington's gene. But when he began to test the probes on DNA from a small Huntington's lineage in Iowa charted by Michael Conneally, a geneticist at Indiana University, he quickly hit unexpected pay dirt. The third probe in the batch seemed to grab onto a marker that showed up consistently in family members with Huntington's—but not in those who did not have the disease. Conneally remembers phoning Wexler with the news. "She let out a scream," he says.

It seemed inconceivable that Gusella could have gotten so lucky so soon; besides, the Iowa family was far too small to clinch the case. So he turned to Wexler and the Venezuelans. She had been giving him samples of blood since her first collecting trip in 1981—she'd send them along with any researcher heading toward the Boston area—so Gusella had plenty to work with. Conneally ran the statistical checks on his computer. One after the other, the samples confirmed the early finding. Through an extraordinary stroke of luck, they had found a marker for the Huntington's gene.

As it turned out, the polymorphism used as a marker came from the short arm of chromosome 4, which meant that the Huntington's gene was there as well. And 96 percent of the time, in each and every lineage, some version of the marker—there are 20 in all—traveled with the Huntington's gene. That meant that if you could determine which version traveled with the gene in each family, you could tell with 96 percent accuracy whether or not a person would develop the disease.

The finding unleashed an ethical storm. In effect it constituted a predictive test for a disease for which there was still no treatment, much less a cure. And it could be used only on people who had living relatives who were both sick and healthy so that the marker could be traced. Before the marker was discovered, 70 percent of people at risk said they would want to be tested for the disease, if such a test were available. Yet in the decade since the marker test became available, only about 13 percent of the at-risk population has been tested.

Ironically, the Wexlers chose as a family not to use the technology. Both Alice and Nancy said that a positive result for either one would devastate all of them. "If you take the test, you have to be prepared to be really depressed," said Nancy. "I've been depressed. I don't like it."

With the marker found, Wexler set her sights on the gene itself. She continued to organize workshops and seek out researchers willing to work on the gene hunt. She traveled from coast to coast, from the United States to the United Kingdom—in between forays to Venezuela, of course—and in early 1984 she and Tobin pulled together the formal Huntington's Disease Collaborative Research Group. At their suggestion, participants in the group agreed to share information during the search and to share the glory when it came to an end. The collaboration pitted a "socialist model"—their group—against several independent laboratories pursuing the gene on their own. They had two main competitors: a lab that had been invited to join but declined, and another that had personality conflicts with people already in the group.

Even within the group, collaboration on this large a scale often created tensions and resentment. "Sometimes someone would charge that one group was withholding a particularly useful marker, or that another was not sharing data as quickly," says Tobin. "Whenever there was a question about whether someone was sharing, I would call the person and say, 'Gee, it would be awfully nice if you brought that material to the meeting.' So it was typical for people to come with little tubes of recombinant DNA that contained a new marker, a new piece of DNA from the suspect region of the chromosome where we thought the Huntington's gene was located. They would distribute it at the beginning of the meeting, and of course then everybody felt better."

Wexler traveled from lab to lab through the 1980s, soothing bruised egos, seeking new talent, and cheering on anyone who was losing momentum. Conneally remembers when he was having trouble finding a place to store cell cultures from the Iowa family on whom the first marker tests had been performed. The only available mutant cell bank, in Camden, New Jersey, took a maximum of three samples from any family. "They were persuaded and cajoled by Nancy to take more," he recalls. "She got it up to 15 individuals. We collected blood from 30 and sent it to them. They couldn't simply throw it away, so they stored it."

Through it all, Wexler returned to Venezuela each year, collecting blood and data until the pedigree spread like polka-dot wallpaper along the corridors outside her office at Columbia University Medical Center in New York, where she had begun lecturing and doing research in 1985. She began to pick up some peculiar patterns, patterns similar to ones Conneally had told her he'd seen in the Iowa family and others. For instance, in some families the gene did not wait until middle age to strike but hit children as young as two. In these juvenile-onset cases, it struck with particular intensity, causing stiffness in addition to the chorea, and death within a decade. And in most of the juvenile cases, the affected parent was the father: indeed, as a Huntington's father continued having children, often successive children would develop the disease earlier and earlier in life.

Wexler particularly remembers one woman in Venezuela who had been sick from the time she was 14. Now she was 21 and she lay dying in a clinic. Her body had wasted away and stiffened to immobility. When Wexler, sitting at her bedside, enfolded the woman in one of her trademark hugs, the woman smiled up at her. "It took a while to develop, but it was a radiant smile," Wexler says. "Yet I knew that chances were, the next time we returned she would be dead." She remembers the frustration. "I felt like I was staring at the answer," she says with quiet intensity. "I knew that locked in that woman's body was the answer to how Huntington's disease works."

Nothing could slow her down in the pursuit of that answer. When she wasn't in Venezuela, or touring labs, or hunting for talent, Wexler joined her father to raise money for the foundation. One of her more successful forays came when she spoke at a dinner hosted by Dennis Shea, a prominent Wall Street financier whose former wife has Huntington's and whose children are thus at risk. In that one night $1 million was raised for the Huntington's fight.

It would take more than money to win this battle, however, and the gene hunters were working with several handicaps. Although Huntington's was the first gene mapped to a specific chromosome through markers, the markers didn't lead directly to the gene, as the researchers had hoped. Indeed, researchers looking for other genes—those searching for the cystic fibrosis gene, for example—were able to use the marker techniques to capture their quarry much more quickly. The Huntington's hunters, though, had trouble finding markers on both sides of the gene, which would help them narrow their search. Furthermore, they had no biochemical clues as to what the gene might actually be doing, so they couldn't search for it in any logical, function-based way. Then, in 1990, they realized they had been looking in the wrong place altogether.

Their efforts had been aimed at a portion of chromosome 4 near the very tip, a confusing region of about 150,000 base pairs that Wexler calls "the Twilight Zone of genetics." As they continued to rule out chunks of DNA, the gene, like a tantalizing mirage, always seemed to lie just out of reach. Finally Gillian Bates of the Imperial Cancer Research Fund in London managed to clone the entire tip of the chromosome—a step that would ensure rapid isolation of the gene. But any optimism generated by that development was quickly squelched when the gene hunters realized the gene wasn't there. Instead, Marcy MacDonald, a senior researcher working with Gusella at Mass General, found evidence that the gene lay 2 million nucleotide links down the DNA chain in the opposite direction. The group had already suspected that this region might hold the gene but had avoided it because it contained about 2.2 million nucleotide pairs and might require a decade or more to sequence. The frustration was enormous, yet they had no choice but to take a breath, switch their focus, and press on.

The end came in far less than a decade. In 1992 several laboratories—including the two not in the collaboration—began focusing on a fairly small portion of the target area and isolating a number of genes, though they had no way of knowing which was the Huntington's culprit. In January of this year MacDonald began sequencing one very large gene that had caught her attention. Close to the spot at which its protein-building instructions begin, MacDonald found a trinucleotide repeat, a kind of broken-record repetition of the three nucleotide bases—cytosine, adenine, guanine—that make up the genetic instruction for the amino acid glutamine. When she and Gusella compared genes from normal and Huntington's chromosomes, it seemed that the number of repeats was always higher in the Huntington's genes. "We said, 'This can't be true. It can't be this easy,'" MacDonald recalls gleefully.

Drawing on a vast pool of DNA samples from 75 different Huntington's families—including some from the United States, Canada, Mexico, China, Japan, Africa, Germany, Italy, France, and, of course, Venezuela—MacDonald and two co-workers went into a frenzied two-week period of lab work. Over and over, they spread fragments of DNA onto treated plates, then ran an electric charge through them. The fragments, showing up as black bands on a white plate, separated by weight as they moved across the plate in response to the charge. The heaviest fragments, which contained the most repeats, traveled the shortest distance; the lightest fragments, containing the fewest repeats, traveled farthest. The researchers were able to tell the number of repeats in each fragment by determining precisely how far it had moved.

In every instance the genes fell into line. In normal individuals there were between 11 and 34 repeats of the glutamine code; Huntington's patients had 37 to 86 repeats. There was no ambiguity, no overlap. In addition, some of Wexler's most perplexing puzzles began to come clear. The researchers found that the youngest victims carried the most repeats, and that the number of repeats tended to expand as the gene was passed from generation to generation. They also found that sperm cells from a man with

Huntington's could range wildly in the number of repeats they carried, though the rest of the man's tissues had but one consistent number. Researchers are now looking into whether, as these men age, they have a higher proportion of sperm with lots of repeats.

After the final experiment, on February 26, the triumphal report was sent to the journal *Cell.* As promised, it was signed simply the Huntington's Disease Collaborative Research Group. And when the participants' names were listed at the bottom, Nancy Wexler's was among them.

The press conferences and parties followed quickly. A month after the paper was published, everyone flew to Dennis Shea's estate in the Florida Keys for a communal sigh of relief. They had been going there once a year since 1987, spending their days working on the beach, and their nights at a bar called Woody's, listening to the hard-driving rock and roll of the house band, Big Dick and the Extenders. This time was no exception. When Big Dick saw the scientists at the bar, he stopped in midsong. The 6-foot-6 singer called to David Housman, who was wearing a T-shirt printed with the likeness of a tuxedo, "Hey, Doc, why don't you come up here and tell the folks what you did!" Self-consciously but with pride, Housman jumped onto the stage and said, "We're molecular biologists, and we've just found the gene for Huntington's disease." Nancy Wexler sat in the back and grinned.

Within days, Wexler was bouncing from Los Angeles to New York to Texas, planning more workshops—first the Philadelphia conference, then a conference on the riddle of the trinucleotide repeats. It is the repeating sequence that now consumes her. How does the stuttery repeat change the normal gene into a killer? After the gene was pinned down, the researchers quickly found that it was expressed in every tissue—yet it seems to devastate only a few cells found deep in the brain. Why? There are some signs that mitochondria, the energy factories in cells, might be harmed by the altered gene product, but how?

The answers to these questions may come from yet another collaborative group Wexler began to put together a few years ago. This one will share the precious supply of several hundred brains and other pathological specimens from Huntington's victims around the world—including the 21-year-old woman from Venezuela, who died just months after Wexler last saw her. The hope is that this tissue might hold new clues to the workings of the disease, and that it might divulge its secrets to researchers willing to work together.

Wexler returned to Venezuela late last winter, just before the *Cell* paper was published. She went mainly to gather new data—the work is never truly done—but also to spread the good news. She walked through a maze of alleys in a shantytown near Maracaibo and saw old friends from her years of research, many of whom were showing signs of chorea. As she waved at them and smiled, she says, she couldn't help but visualize on their faces the broken-record repeat from the plates back in the laboratories, like the shadows cast by a venetian blind. Everywhere she looked, the trinucleotide stutter looked back, dizzying in its persistent mystery.

"I had spent so many years being so curious about what it was, studying all these people whose bodies contained the mystery," Wexler says. "And suddenly it was superimposed on them, almost like a silk screen. It was an image without words, saying, 'Here's the answer. And here's another question.'"

THE HUMAN GENOME PROJECT[4]

Geneticists Raymond White and Francis Collins were beaming when they stood before television cameras last July. They had just discovered the gene that causes neurofibromatosis, the famed "elephant man" disease.

The smiles broadened a month later when workers in White's lab said they had learned that the newly found gene was identical with one that seems to play a role in many human cancers. At one swoop, genetic research had given an insight not only into a brutally disfiguring disease but also into one of the leading killers.

The TV cameras rolled again the following month at the National Cancer Institute in Bethesda, Md., when Steven Rosenberg and his collaborators announced they finally had permission to try gene therapy to cure a usually fatal disease. Within hours, they

[4]Reprint of article "Genome" by Edward Edelson from *Popular Science Magazine* v 239 pp58–63+ Jl '91. Copyright © 1993 by Times Mirror Magazines INC. Distributed L.A. Times Syndicate. Reprinted with permission.

were injecting genetically altered cells into a five-year-old girl who had been born with a seriously flawed immune system. [Editor's note: The experiment referred to was the first gene-therapy trial carried out on a human patient; it is discussed more fully in "Designer Genes," the sixth article in this section.]

And just a month later, researchers said they had done test-tube studies in which they had corrected the cellular defect that causes cystic fibrosis, the most common genetic disease among white Americans. The cystic fibrosis gene had been discovered only a year earlier, and now research opened the possibility of a near-normal life for patients whose survival once was counted in terms of a few painful years. Weeks later, Rosenberg and his collaborators started a second gene therapy trial, this one for patients with terminal malignant melanoma, a particularly vicious skin cancer.

All these advances were based on one achievement: the identification and isolation of specific human genes. Impressive as the results seem now, leaders of America's genetic research community say that they are comparable to Henry Ford making cars by hand, one at a time. Geneticists have an assembly-line plan they say can revolutionize not only genetic science but all of medicine. It's called the human genome project, and its goal is nothing less than the mapping and isolation of all the genes in the human cell.

It's inefficient to continue today's fragmented approach, these geneticists say. Collins, at the University of Michigan [Editor's note: Francis Collins was appointed the head of the Human Genome Project in 1993. His ideas on the future of the project are discussed in the following article.] and White, at the University of Utah, had to work for two years to find the neurofibromatosis gene. The estimated cost of discovering the cystic fibrosis gene is tens of millions of dollars. Genetic discoveries will be made a lot faster and cheaper once the genome project gets rolling, its proponents say—and it will have many other major benefits as well.

"The Human Genome Initiative will give us two things," says Leroy Hood, a world leader in genetic research at the California Institute of Technology in Pasadena. "It should promote technological development of instrumentation the likes of which we've never seen before, and it's going to give us this encyclopedia of life. That isn't an experiment. It's a resource. It's going to be in a computer, and it's going to let biologists do all sorts of fancy things thereafter."

There's a glimpse of Hood's dream of the future in the laboratory of Glen Evans, a molecular biologist at the Salk Institute for Biological Studies in La Jolla, Calif. Evans is one of many bright young scientists bursting with ideas about how to map the genome, the term for all the genetic material of a species. In his laboratory, you can watch a machine, a simpler version of a laboratory robot that Hood invented, go through a set of tedious movements that once would have been done by a technician. It can literally build a gene, unit by unit, by following computer instructions.

"We just type in what the sequence is, make sure all our bottles are filled, and come back three hours later," Evans says. "This is the cheap version. It costs about $12,000, as opposed to the robot, which is about $30,000."

Evans quoted a few other prices: a microscope that lets him see genes directly, $20,000; a micromanipulator that lets him slice out a section of genetic material, $15,000. But his dream machine —another instrument that came out of Hood's laboratory—is one that can read the gene sequences automatically. "It's incredibly useful, but we don't have it. It costs $100,000. When we have to determine a sequence now, we do everything by hand. If I had an automated sequencer, I could do the same thing but in a much more efficient way." The human genome project would give him —and others like him—those tools.

Hood and Evans are fervent believers in biology's first megaproject, an effort that will cost about $3 billion and run for at least 15 years. But there are equally fervent opponents. The human genome project has aroused strong emotions on both sides, and those emotions are growing stronger as the program picks up speed.

Feelings range from the awe expressed in the description of the project by Walter Gilbert of Harvard University, a Nobel laureate in genetics, as the "Holy Grail" of biology to the curt dismissal of the program as "mediocre science" by Martin Rechsteiner, a biochemist at the University of Utah who is spearheading opposition to the plan. There are also worries that the project could lead to a twisted 21st-century version of Nazi eugenics, in which people are stigmatized for the genes they carry. Counterbalancing those fears are hopes that the genome project could bring medicine unmatched abilities to ease and improve the human condition.

Hood and other geneticists say the injection of some serious sums of money now will bring rewards of incalculable importance and benefit to humanity.

"To get the human genome is priceless," says Norton Zinder of Rockefeller University in New York, who chairs the National Institutes of Health (NIH) advisory committee on the project. "It's worth more than a lousy aircraft carrier, which is five billion dollars. Just to have the genome—what can you compare it with?"

"Why do we want to spend the money?" asks James Watson, the Nobel laureate who is heading the project [Editor's note: James Watson resigned this post in April 1992.]. "The answer is there is no better way to understand anything in biology than to go straight for the gene that's involved. We have an example of that in cancer. Only when it became possible to isolate the genes that cause cancer could we understand what went wrong."

Behind those statements is a belief that genetics is ready for a mighty leap forward—that biomedical technology is building the tools that make possible a complete understanding of our genetic inheritance at a reasonable price.

It's a peculiar territory they're exploring: a ribbon of a molecule called DNA—deoxyribonucleic acid—two yards long and a few microns wide. You've got that much DNA in almost every cell of your body, divided into 46 units called chromosomes. You inherit them in pairs: 23 from your mother, 23 from your father. Every time a cell divides, the two daughter cells get a full complement of chromosomes.

The DNA that makes up those chromosomes consists of two matching strands coiled in a helix—a discovery that won Watson his Nobel Prize. The immensely long DNA molecule is made up of subunits called nucleotides, or bases. There are just four different bases in DNA, named cytosine, guanine, adenine, and thymine, usually referred to by their initials. In the two DNA strands of a DNA molecule, a C of one strand is matched with a G in the other and an A is matched with a T. . .

There are some three billion bases in the human genome; somewhere in that long thread are sequences of bases that code for all the proteins that make life possible. Each sequence is called a gene, and the best estimate is that we carry from 50,000 to 100,000 genes in each of our cells.

Molecular biologists have made incredible advances in manipulating this system. They can snip the DNA molecule to cut out a gene and put it into bacteria, where it busily turns out its protein. Voilà, the genetic engineering industry, which has made many a molecular biologist a millionaire.

Using the machine invented by Hood and his colleagues, biologists can make their own strands of DNA, working either with genes or probes to find matching strands. Yet with all their advances, they've barely begun to get the kind of information about the human genome that they want.

"The catalog of all the human genes we know anything about has just gone to press with its ninth edition," says Victor McKusick of Johns Hopkins University in Baltimore, the man who keeps the catalog. "The number is 5,000. There's a long gap there between 5,000 and 50,000. And the total number of genes mapped to chromosomes is only 1,900."

Most of the genes mapped so far are those that cause disease. And mapping isn't cheap. "Obviously, it's not cost effective to do genes one at a time," Zinder says. "If we ever get the whole map done, any time you're interested in a gene you could reach in and get it."

When Zinder says "the whole map," he's talking about two separate things, a physical map and a genetic map. The two maps give you different kinds of information about the genome, are drawn by strikingly different methods, and require different vocabularies.

A genetic map locates genes in relation to one another on a chromosome: Gene A comes before gene B comes before gene C. Biologists have been mapping genes—first to individual chromosomes, then to general regions on chromosomes—for half a century.

A basic technique for making genetic maps is based on nature's sloppiness. When a cell divides, the two strands of DNA in each chromosome separate so they can duplicate themselves. That separation is often flawed. The threads of DNA keep breaking and recombining with other threads. Thus you can tell how close together two genes are by noting how often they become separated by this sort of recombination.

A newer way to map a gene makes use of restriction enzymes, which clip DNA at places where the enzymes recognize specific base sequences. (Bacteria use restriction enzymes to attack invading viruses. Biologists have isolated a bunch of bacteria as tools for chopping DNA into useful segments.) Their use in gene-finding comes from the fact that DNA sequences vary widely from person to person. If you turn the restriction enzyme loose on a chromosome, you'll get a bunch of fragments that biologists call, with their usual indifference to language, restriction fragment length polymorphisms—RFLPs, or riflips. Riflip patterns differ from person to person because there are individual differences in DNA sequences. Use the restriction enzymes on chromosomes from a husband and wife, and you'll get two different riflip patterns.

The way you use riflips for genetic mapping is to study a big family afflicted with a genetic disease and find a specific riflip pattern that's present in family members who have the disease but not in the disease-free members. That grueling and expensive process has produced some notable successes: location of the genes for Huntington's disease—the malady that killed Woody Guthrie—for muscular dystrophy, and for cystic fibrosis, for example. But to do this kind of mapping you need not only the right family but also appropriate "markers," an identifiable physical location on a chromosome—a gene, a riflip, a restriction-enzyme cutting site. The dream of gene mappers is to have a regularly spaced set of markers at close intervals. [Editor's note: The first article in this section provides an account of how scientists developed this technique.]

A physical map, by contrast, gives you the structure of the DNA itself. You make a physical map by chopping a chromosome into pieces and putting those pieces into an ordered sequence by identifying their overlapping ends. Physical maps come in various scales, depending on the size of the pieces.

A physical map consists of a number of fragments of DNA that have been cloned and put into a suitable living organism. The pieces are named contigs, because they are contiguous segments of DNA. One favorite place to store contigs is in cosmids, laboratory-modified viruses. A cosmid can hold a DNA fragment approximately 40,000 bases long. A newer storage site is the YAC, yeast artificial chromosome, developed by Maynard Olson at Washington University in St. Louis. A YAC holds 400,000 base segments, so an entire library of human contigs can be stored in approximately 60,000 yeast cells.

The ultimate physical map is the complete DNA sequence of all three billion bases in the human genome. That goal is furthest away. So far, less than one percent of the human genome has been sequenced, and the work is not advancing at breakneck speed.

Genetic map, physical map, sequence—these don't complete

the list of the genome initiative's goals. As a kind of prelude to tackling the three billion bases of the human genome, geneticists are now trying to sequence the genomes of less complex creatures running to a mere several million bases each. Among them are the mouse, a yeast, a bacterium, the fruit fly and a roundworm, all well-established as laboratory models.

The organizational structure that's going to make all this possible is as complex as the project's goals. There is an Office of Human Genome Research in the National Institutes of Health, which is supposed to supply most of the money. There is a corresponding organization in the Department of Energy, which has carved out its own set of targets. There is a Human Genome Organization, HUGO, started to enlist international support and run separately from any government office (but led by many of the same people who are working on the U.S. efforts). Things could get even more complex as other nations join in.

One such country is the Soviet Union. Despite reservations over the cost of the project, the State Committee for Science and Technology approved a human genome research effort in 1989, one of fourteen scientific-technical programs to receive priority in the Soviet Union. It was organized by the Institute of Molecular Biology of the Academy of Sciences and was initially funded with 25 million rubles—about $4 million at the official rate of exchange.

Evans of the Salk Institute, whose laboratory was one of the first to get money in the initial round of genome funding, is pushing ahead with innovative ideas for physical mapping. For example, he's made a two-dimensional array of cosmids that enables a computer program to find neighboring contigs more easily than before.

"We've taken that idea to its logical extreme," he says. "We thought that instead of using a two-dimensional array, we could use a cube, a three-dimensional array, and see where the planes intersect. Then we said we could compress it even more by doing a set of cubes, which would be a hypercube. We've written an article called 'Genome Mapping in Multidimensional Spacing.' It turns out the limit is a five-dimensional array."

Physical mapping and genetic mapping can go together, Evans adds. He's using probes that detect DNA sequences known to be associated with genes to help find genes on contigs. He's come up with more than a dozen candidates that are being investigated.

Another bright young researcher, David Cox at the University of San Francisco, also has a new strategy for genetic mapping. He calls it radiation hybrid mapping: Put copies of a single chromosome in hamster cells, bombard them with X-rays to break up the chromosome, and then study the resulting fragments. "We use 8,000 rads of X-rays, and one of our map units is 50,000 base pairs," Cox explains. "That gets you in a range where you can actually clone the pieces of DNA." You can tell the closeness of two genes by seeing how often they are found on the same chromosome fragment.

Both Evans and Cox illustrate how the genome project is bringing together highly competitive geneticists. Cox has organized a concerted effort to map chromosome 21. The 30 or so major groups working on that chromosome have agreed to pool their data to develop a common set of landmarks for it. And Evans is one of six researchers who are getting a complete set of Olson's 60,000 YACs, developed at enormous effort. Both developments have biologists shaking their heads in delighted amazement because they break the tradition of never letting your neighbor know what you're doing until it's absolutely necessary.

An equally amazing development was an agreement reached in May 1990 by all the major gene mappers to pool their data for a definitive map, with 150 markers evenly and closely spaced across all the chromosomes. "Jim Watson and Maynard Olson deserve the Nobel Peace Prize for that," says one biologist familiar with the antagonisms and rivalries among gene mappers.

Another potential sore point, the participation of the Department of Energy, also seems to be yielding to the spirit of good will. With the same candor its predecessor agency brought to the atomic bomb, the DOE has parceled out three human chromosomes to three national laboratories: number 21 to Lawrence Berkeley in California, 19 to Lawrence Livermore in California, and 16 to Los Alamos in New Mexico. Bureaucratically, it makes sense. With the apparent end of the cold war, the DOE's national laboratories are looking for a new mission. But many scientists—including some in the genome project—aren't sure it will make scientific sense because the DOE doesn't have a great background in genetic research.

"We're very much in the setting-up stage," says Charles Cantor, a world-class geneticist who works at Lawrence Berkeley. Take physical mapping, he says. There's the bottom-up method, where you break a chromosome into a zillion pieces and try to put them in order; and there's the top-down method, where you break the chromosome into a few pieces, then break those into small pieces and put them in order before assembling the big pieces.

"What we're trying to set in motion is a couple of second-generation strategies," Cantor says. "In one we'll try to go from large fragments of DNA directly to clonable smaller pieces that are preassigned to the big pieces. We'll reduce the problem of making contigs to forty small problems rather than one gigantic one. The other approach we're using is to order and fingerprint not random fragments but two carefully chosen specific libraries that overlap just at their edges. Pilot studies are being done, and we'll be in large-scale implementation in six months."

But the all-out enthusiasm of most geneticists for the genome project isn't shared by many people outside their field. Theodore Friedmann, a physician at the University of California at San Diego who is working on gene therapy, says the genome project won't give him any immediate help. "For the time being, we need to solve the technical problems we have with existing models—how to insert genes into cells, how to ensure they are expressed stably without injuring the cell."

Arno Motulsky of the University of Washington in Seattle, one of the nation's leading figures in genetic medicine, was more positive—with one notable reservation. "The more we know, the more we can do testing that will help people," he says. "We wouldn't understand a thing about genetic disease if we didn't have the basic science behind it. The field is ready for the mapping of the human genome. But I'm not enthused over the sequencing of the genome."

Figuring out the complete sequence is the most controversial part of the project. One big reason is that most of the human genome consists of nonsense—long, repeating segments of DNA that don't seem to say anything at all. Only a small portion of the human genome, maybe two to five percent, encodes for genes. Many scientists don't see much sense in spending money to read those nonsense sequences. For example, Robert Weinberg of the Whitehead Institute in Cambridge, Mass., whose work with cancer genes is world class, is on record as opposing the exploration of what he calls "a vast sea of drivel" in human DNA.

The organizers of the genome initiative say they're not going to do that kind of exploring for a while—for economic reasons, they claim. Sequencing costs a lot. In Hood's lab, it costs $1 a base.

In most other places, it costs closer to $10 a base. The official plan now is to delay mass-production sequencing until the cost is brought down 10- to 100-fold, to 10 cents a base or less. Scientists say that they will assess sequencing progress over the next five years before they decide on an all-out assault on the sequence.

The leaders are convinced they'll get their total sequence sooner or later and that it will be worth it. "There may be reasons we don't know for keeping the extra DNA," says Zinder of Rockefeller University. "We can call ninety-five percent of it junk, but we have to think of it differently. Would it work better if the DNA encoding genes were shrunk down to one percent? Are there structural reasons for embedding DNA within DNA? We haven't the vaguest idea. Only when we have lots of it sequenced will we be able to figure out if there is a reason."

Doubts have also been expressed about the scientific value of the project. "In principle, their catalog would make my work easier," says Robert Tjian of the University of California at Berkeley, who works on the proteins that regulate gene expression. "My worry is it's going to be a long time before these guys get to the point where they're providing us with valuable information."

Economics looms behind much of the criticism. These are tough times in biological science, especially for the young researchers who are the hope of the field. Fewer and fewer are getting the first grants they need.

"American biological science is in jeopardy because of the current grant situation," Rechsteiner of the University of Utah says. "Diversions of any kind to a big project like this with little training for young scientists are unwarranted."

Zinder and other geneticists say the real money problems have other causes—the vast increase in spending on AIDS, for example—and that the genome project isn't taking funds from anyone because it's all new money. That argument doesn't impress Rechsteiner. "If you were going to add $200 million to the NIH budget, I would argue there are better ways to spend the money scientifically," he says.

Then there are the ethical worries about misuse of genetic knowledge. Friedmann recalls the turmoil that came with mass screening for the sickle cell gene among American blacks in the 1970s, when many sickle cell carriers unjustly experienced problems with employment and insurance because of the costs incurred by victims of sickle cell anemia. "With long-range health prognostication possible through genetic analysis, tensions between the interests of individuals and those of employers and insurers will become increasingly severe," he says.

Foreseeing those objections, the genome project is allocating two percent of its budget for studies of ethical, legal, and social considerations. An advisory committee already has laid out a coordinated program of study and education to deal with those problems. One of the people on that committee was Jonathan Beckwith, a molecular biologist at Harvard who is a leader of Science for the People, a determinedly anti-establishment group that has often protested about neglect of human values in science.

"The funny thing is that in my mind, the importance of the genome project is exaggerated in terms of social consequences," Beckwith says. "A lot of the things I'm worrying about are happening independently of the genome project. In that sense I'm pleased by the project, because it's the first time that funding has been provided to think about social consequences at the same time that a project is funded. I think it's a good precedent."

An entirely different problem that concerns people in the project is data storage and management. It's not so much putting the data into computers—"Anyone can store three billion bits of information," one scientist says—as developing methods to make it usable. Data bases are being set up in several places, including Johns Hopkins University and Los Alamos National Laboratory. But much of the informational work for the project is being done in Bethesda, Md., at the National Library of Medicine's Center for Biomedical Technology—which, curiously enough, isn't an official part of the project.

"You can't just waft it into the computer someplace and leave it there," says David Lipman, director of the center. "You have to come up with new hybrid ideas. You have to understand at the intuitive level what biologists want to ask. In past searches of the data, just experts were doing it. Now every molecular biologist has to be able to do sophisticated searches on his own, or you don't get the maximum out of the data. Some people want to throw up their arms and say the computer problems are going to be impossible. I don't think that's true. I'm confident it's doable."

One measure of the difficulty is in the size of the first genome project data base being prepared by the center. It will be too big to go on one CD-ROM, which can handle an ordinary encyclopedia with no problem, Lipman says. "We'll be making CD-ROMs with a variety of information on them, but we'll have to set up the data in various ways."

Chips for Genes

Hood is also working on genome computer technology. He's already got one microchip, derived from space science, that's superfast at searching through genetic sequences to find matching patterns. He and researchers at the University of Southern California and the Jet Propulsion Laboratory in Pasadena, Calif., have just finished a second chip, being made by Hewlett-Packard, that is "ideal for searching for unknown patterns." Chips like those could pick genes out of a sea of nonsense DNA.

The genome initiative is the kind of challenge America has traditionally accepted. Says Hood: "If we aren't rich enough as a country and intellectually daring enough to do this, we're in disastrous shape for the future."

TAKING STOCK OF THE GENOME PROJECT[5]

Francis Collins has never been one to avoid tough jobs. Collins, who was lured to the National Institutes of Health (NIH) last spring to head the National Center for Human Genome Research, has hunted down a string of elusive genes. He tackled cystic fibrosis, joining the team that eventually pulled out the faulty gene and has since been working on a therapy. He co-identified the gene that causes neurofibromatosis 1, and was part of the collaborative group that finally tracked down the gene involved in Huntington's disease earlier this year. Collins and others are now pursuing a breast cancer susceptibility gene—a discovery that could have vast medical, social, and ethical implications.

But in his new job, Collins is confronting a bigger challenge

[5]Reprint of article by Leslie Roberts from *Science* v 262 pp20–22+ O 1 '93.

than he's ever faced before. He is taking over the reins of the genome project, formerly headed by Nobel laureate and double helix codiscoverer James Watson, at a critical time. The international program, which is jointly funded here by the Department of Energy (DOE) and NIH, needs a considerable infusion of new dollars if it is to meet its ambitious goal of determining the sequence of all 3 billion base pairs of the human genome by 2005. Yet the project is already getting about $165 million a year, and money for science is scarce. "When you see all the programs that have lots of support taking cuts, it is very difficult to argue successfully for a ramp up of an innovative new program," concedes Collins. "It is a real problem."

Yet funding is only one of myriad issues Collins faces. He and numerous advisers in the genome community have just finished an intensive review of the first few years of the program, taking stock both of its successes and its failures. Based on those findings, they have crafted a new 5-year plan to guide the project through 1998. In some areas, there seems to be a remarkable scientific consensus on where the project ought to go. In particular, everyone agrees there should be a concerted push on DNA sequencing because a lack of new technologies points to a major bottleneck in the years ahead. Essential, too, is increased emphasis on new software and hardware for collecting, disseminating, and analyzing the sequence data. Also in vogue is a shift away from the chromosome as the unit of analysis to a new focus on smaller regions of biological interest—a change that will influence which genome centers survive and which new ones get funded.

On other issues, there's less consensus, however. Researchers are divided over how much effort the project should put into tracking down genes, for example. A program devoted to the ethical, legal, and social implications of the project, known as ELSI, has come under criticism from disgruntled scientists, who want a greater push on public education, and from others who think ELSI should be stepping boldly into policy making. Then there's the ongoing dispute over the propriety of patenting gene fragments.

So far, however, Collins seems to have the support of the genome community, which he will certainly need. Collins, too, apparently has the support of his bosses in the Department of Health and Human Services (HHS), where a proposal is now wending its way up to create another institute for NIH—a National Institute of Genomics and Medical Genetics, with Collins as its director.

First the Good News

By one key measure, Collins has inherited a project in good shape: In terms of meeting its first priority—developing genetic linkage maps of the human genome and those of several model organisms—the project is coming in ahead of schedule and under cost. "We have excellent genetic maps of mouse and human even faster than expected," says Maynard Olson of the University of Washington. Olson attributes much of the success to Jean Weissenbach and colleagues at the Centre d'Etude de Polymorphisme Humain (CEPH) in Paris. Genetic linkage maps, which consist of a series of signposts—usually highly variable pieces of DNA—arrayed along the chromosome, are particularly useful for finding the rough location of disease genes.

A second type of genome map, the physical map, if not ahead of schedule is at least on target. Physical maps are actual assemblages of DNA clones, lined up in the same order as they appear on the chromosome. The ultimate goal is to go from these maps to pulling out genes and sequencing them. At this stage, however, the resolution of these maps is less than ideal. The markers are spaced on average every 300 kilobases, as opposed to the original goal of a 100-kilobase resolution map by 1995.

Even so, the increasingly sophisticated maps and resources, such as DOE's chromosome-specific collections of clones, have speeded the isolation of genes involved in numerous diseases, including Fragile X, Huntington's, and colon cancer. Studies of these genes have, in turn, revealed fascinating genetic mechanisms, such as the trinucleotide repeat mutations that lie at the heart of Fragile X, myotonic dystrophy, Huntington's, and who knows how many other diseases.

Mortgaging the Future

But progress on the maps has come at a cost. While mapping is ahead of the schedule originally set in 1991 in the first 5-year plan, sequencing lags behind. Although sequencing speed has risen over the past few years and the cost per base pair has dropped, Collins and others say a 100-fold improvement in speed is still needed if the project is to meet its goal of knocking off the entire human genome by 2005. Indeed, it was partly concerns about the sluggish progress on sequencing that prompted Collins, David Galas, who, until he recently left for Darwin Molecular Technologies in Seattle, oversaw the genome project for DOE, and numerous advisers to revisit the 5-year goals over the spring and summer. The roadblock in sequencing is now money, says Collins. "Good ideas are going begging," he says.

The problem is that the budget has not increased as fast as the project's creators recommended. When biology's first megaproject was planned, a committee of the National Research Council in 1988 concluded that it would take 15 years and cost about $3 billion, or $200 million a year, to pull it off. Although those numbers have withstood repeated scrutiny, asserts Collins, the $200 million has failed to materialize. The combined NIH and DOE budget remained at roughly $165 million from 1992 to 1993, when, adjusted for inflation, it should be at $219 million, says Collins. "We are now being asked to do the project at 75% of the funding that the NRC said it should cost." The upshot, he says, is that most of the money went into mapping, and the "revolutionary sequencing techniques envisioned earlier simply have not materialized. We have mortgaged part of our future."

The new plan calls for $100 million a year exclusively for sequencing technologies, thereby bringing the total budget up to the $200 million equivalent originally recommended. What happens if they don't get it? "The simple answer," says Collins, "is that we are probably not going to be able to make that timetable." And he predicts the consequences would be grim, both in terms of delayed medical benefits and a loss of U.S. biotechnology competitiveness.

What's more, even with the full funding, meeting the sequencing goal will still be a "stretch," concedes Collins. He and others predict that the job will probably have to be done with incremental improvements in today's sequencing technology, based on gel electrophoresis, rather than with glitzy new approaches such as mass spectrometry or atomic force microscopy. What's needed now, Collins and others agree, is automation that will dramatically lower the amount of labor needed to sequence and thereby cut the costs—something that "the genome project has failed to get a grip on," asserts Olson. Until now, people have largely focused on automating individual steps of the sequencing process, says Collins—for instance, building better sequencing machines, or robots for DNA preparation, or software that can analyze and assemble clones in order. Now the focus is shifting toward automating all the steps as a unit so that no one step is rate limiting.

But that poses a tough question about the sequencing budget. We could put all of our eggs into automating current sequencing methods, which we know will work, says Collins. But what then, he asks, about the "blue sky revolutionary ideas" that don't get funded because of the budget crunch—and that could make all the difference?

Despite the slow progress, there is little sentiment at this stage for abandoning the goal of all-out sequencing, Collins says. The biological insights emerging from the few large scale sequencing efforts, such as those on the nematode *Caenaorhabditis elegans,* are just too alluring. Comparative analyses have revealed, for example, a remarkable similarity in genes shared across species.

But some thought is being given to a shortcut called one-pass sequencing. The original plan calls for sequencing the whole genome several times, to ensure an error rate of 0.001%. "Suppose we try one-pass coverage with 1% error rate but it only costs one-tenth as much?" asks Collins. The idea, then, would be to return to the really interesting regions and sequence them again.

Recruiting New Bodies

Aside from cold cash, the sequencing effort will need more warm bodies if it is going to be done on time. Additional groups will have to get involved in large-scale projects, says Collins, tackling a sizable chunk of a chromosome, say, a megabase a year. Right now, Collins can count on one hand the groups that have that capability. One way to recruit sequencers and convince them to scale up is to give them some interesting biology to work on. And that meshes nicely with another change in the new 5-year plan.

Most of the existing genome centers were built around analyzing particular chromosomes. But now there's a shift away from chromosomes to focus instead on regions of biological interest—particularly regions several megabases long that seem to have functional, structural, and evolutionary significance. "It's biology that should dictate the size" of the unit sequenced, says Olson. And to him, that is good news, because it means the genome project will be "more biologically driven, with more room for creativity, and that the centers won't have to be so huge."

Putting Genes on the Map

Another major shift in emphasis under Collins, perhaps less universally embraced, is a goal of placing the 100,000 human genes on the maps. Gene identification was always an implicit goal of the project, insists Collins, though it was never stated explicitly, perhaps because of its difficulty. Now several new techniques make gene finding easier, he says, and the annotated map will be far more useful in helping investigators identify disease genes.

Articulation of this goal is also sending an explicit message to the private sector, where enormous efforts—"probably more than we know about," asserts Collins—are under way to partially sequence all the expressed genes or cDNAs. Pending a decision of the patentability of these gene fragments, most of these sequences are being kept secret, says Collins, and the genes are not being put on the map.

But this new emphasis on gene identification is raising questions about what the goal of the genome project really is. As first envisioned, it was to build the infrastructure—the maps and tools—to prepare for the biology of the 21st century, leaving the gene discovery for others. Will Collins, the avid gene hunter, shift the focus too much in the direction of looking for disease genes, especially when technology development is suffering? That possibility worries Leroy Hood, head of the molecular biotechnology department at the University of Washington and also a large-scale sequencer. Hood sees no fundamental role for disease gene hunting in the project, because "there is simply not enough money to go around."

Collins actually agrees, asserting that "building the infrastructure is still our first priority," and pointing out that he means annotating the map with all genes, not just disease genes. But, he adds, "the reason the public pays and is excited—well, disease genes are at the top of the list. We can't take on the entire field of finding genes, but I will be pleased if the project catalyzes it along the way." Collins suspects that some of these concerns may reflect apprehensions about the intramural program he is launching at NIH, which will have a decided focus on disease genes and indeed gene therapy. But, he insists, "the creation of an intramural program with a strong applied focus does not change the extramural program."

ELSI at a Crossroads

From the start, Watson and others realized that the information garnered from the genome project could be misused—in denying health insurance, for instance. For that reason he set up an ethics program and promised that it would receive at least 3% of genome project funds. It now receives about 5%. But now, having spent 4 years and $20 million, the ELSI program is at a critical juncture, with numerous critics wondering what it has produced. Asked Olson at a meeting this summer: "Why don't we have any visible progress toward a federal genetics privacy law 3 years into the program?"

Most of what ELSI has done has been to define the high-priority issues that require urgent attention, both through research grants, such as pilot studies on screening for cystic fibrosis, and, most visibly, a series of academic meetings. It is these meetings—where often the same cast of characters debate the same issues—that have taken most of the heat. Some bench scientists are openly fed up. "We've had enough of this Hastings Center stuff," says one. This summer several advisers to NIH and DOE's genome projects complained that ELSI was too divorced from the science and that it was time for them to quit talking and start doing something, though opinion is divided on whether it should be active public education or policy making or both. Even Collins, a staunch supporter, concedes that "people are tired of another venue of defining the issues. It is time to move on and produce some general policy recommendations."

But does ELSI have either the clout or the independence to develop policy recommendations that anyone will listen to? In a report last year, the House Committee on Government Operations described ELSI as well intentioned but too low in the bureaucracy to be effective at setting policy. The committee recommended that an independent body be created to review the ethical, legal, and social implications of the genome project. An upcoming report from the Office of Technology Assessment is also expected to support the notion of an independent bioethics commission.

While Collins maintains that ELSI is independent enough to serve as watchdog, he also says he has no problems with a new commission, provided its budget doesn't come out of ELSI funds. ELSI would still have an indispensable role to play in catalyzing research on ethical issues—particularly on the proper introduction of genetic screening for such common diseases as breast or colon cancer, he says. Other critical roles are educating the public about genetics and its implications and developing policy options for any new commission to consider.

One of the most pressing issues Collins will confront is what type of genetic information can be patented. Hood characterizes this as "the most confused and potentially challenging issue we face. Whether we like it or not, Craig Venter opened up a whole series of questions that are legitimate." Hood is referring to former NIH scientist Craig Venter, now at The Institute for Genomic Research outside Washington, D.C., who sequenced thousands of gene fragments of unknown function. NIH kicked off a furor by applying for patents on the fragments. When the Patent Office rejected the application last year, HHS appealed that decision. What happens next will depend, in large part, on how aggressively incoming NIH Director Harold Varmus pushes the appeal. Collins is hoping the NIH patent will soon be disallowed. Until then, he says, massive cDNA sequencing is proceeding apace in the private sector and those sequences are not being made public. "It is bad for the project."

But even a resolution of this one patent application will not address the other questions Venter's approach has raised. While there seems to be near-universal agreement that gene snippets should not be patentable, what about entire genes, or entire regions, like the T cell locus, which encodes genes involved in the immune response, ponders Hood. "If we complete the sequence of the T cell locus, should we be allowed to patent all the genes even without knowing their function?"

Tricky questions, too, are arising from the myriad of new startup companies based on the genome project, which Collins sees as a sign that the genome project is indeed succeeding. But such success is "twoedged," he concedes, as many of the principals are heavily involved in NIH-funded genome centers. To name a few: David Cox and Rick Myers of Stanford, along with Dennis Drayna, formerly of Genentech, just founded Mercator Genetics Inc., in Menlo Park, and Eric Lander of the Whitehead Institute and Daniel Cohen of CEPH are "founding scientific advisers" at Millennium in Cambridge. The obvious concern is that some investigators might use the resources developed with public funds for their own proprietary interest. Yet another question is how will the genome project receive impartial advice when nearly everyone has a financial stake in it? "We must all be willing to sit under a hot light," says Cox. And Collins, who rid himself of all commercial ties to take this job, vows to keep a tight watch.

Once Collins settles into his new job, he says that he plans to spend about one-third of his time doing science in the intramural lab. For now, he has his hands full moving his lab from Ann Arbor and recruiting some of the nation's top geneticists to join him at NIH. As he dashes from city to city, lobbying for a budget increase on Capitol Hill one day, attending a thesis defense in Ann Arbor on another, Collins seems energized, not cowed, by the challenges that face him. "I like intensity," says Collins, who insists he has yet to regret, even for a moment, his decision to take this job.

THE GENETICS REVOLUTION: DESIGNER GENES[6]

On September 14, 1990, a four-year-old Cleveland-area girl made medical history by becoming the first human to be "operated on" in a gene therapy experiment. A defect in one of her genes rendered the cells of her immune system unable to combat infections. This extremely rare hereditary disorder is known as adenosine deaminase (ADA) deficiency—nearly identical to the condition that afflicted "David," the highly publicized boy who lived his life isolated in a plastic bubble and died in 1984.

In an effort to give this girl a normal life, doctors at the National Institutes of Health (NIH) removed some of her blood cells, fortified them with a healthy copy of the faulty gene, and then reinfused the now-potent blood cells back into her circulatory system.

The therapy worked even better than the researchers had hoped. In less than a year, the girl's immune system was functioning well enough for her to enroll in kindergarten. A nine-year-old girl with ADA who subsequently received the same treatment now attends elementary school.

In the short time since it was first tried, gene therapy has

[6]Reprint of article by Beverly Merz from *American Health* v 12 pp46–54+ Mr '93. Copyright © 1993 by Beverly Merz. Reprinted with permission.

soared from a theoretical curiosity to one of the hottest areas of medical research. The NIH Recombinant DNA Advisory Committee—which, along with the Food and Drug Administration (FDA), rules on all genetic therapy requests—has given several additional research teams the go-ahead for gene therapy.

These pioneering efforts are targeting conditions for which treatment options are extremely limited, including brain cancer, advanced melanoma, cystic fibrosis and a hereditary condition that results in extremely high cholesterol levels in the blood. In coming years, genetic fixes will probably be attempted for many more disorders, including breast cancer, colon cancer and diabetes.

For some people, gene therapy conjures up disturbing images of sperm and eggs manipulated to yield tailor-made human breeds enhanced with traits such as superior intelligence. But experimenters seem determined to steer clear of any such tampering with reproductive cells. Instead, current gene therapy is more akin to an organ transplant, with its impact limited to the patient's lifetime.

A key breakthrough in making gene therapy possible occurred 20 years ago, when researchers succeeded in snipping genes from human cells and inserting them into rapidly multiplying microbes such as bacteria and yeast. With that development—variously known as gene splicing, recombinant deoxyribonucleic acid (DNA) technology or gene cloning—researchers wishing to study a particular gene could make millions of copies, or clones, of it in a few days. And since genes direct the production of proteins, genetically engineered yeasts or bacteria could become tiny factories for churning out large quantities of exceedingly rare and valuable human proteins. Today human insulin, human growth hormone and 15 other genetically engineered proteins are being marketed as drugs in the U.S., and over the next two years the FDA is expected to approve many more such drugs, courtesy of bacteria and fungi.

Gene therapy usually involves splicing human genes into much smaller microbes, the viruses, which multiply by inserting their genetic materials into host cells. Taking advantage of viruses' skill in infecting cells, researchers turn them into delivery vans for carrying healthy human genes into cells that have a defective copy of a gene or lack it entirely.

Genes, the basic units of heredity, are linked together in long chains to form the chromosomes in a cell's nucleus. Between

50,000 and 100,000 genes are packed into the 23 pairs of chromosomes in a human cell. Each gene directs the production of one protein; collectively, they provide all the instructions needed to make a human being. The genes and chromosomes are made of DNA, the self-reproducing master molecule of life. . . .

So far, gene therapy has mainly targeted those genes that produce specialized proteins called enzymes, which regulate the myriad processes that occur within cells. A defective gene can result in a faulty enzyme or no enzyme at all. In the cases of the two girls described earlier, a gene defect deprived their lymphocytes (a type of white blood cell) of ADA, an enzyme vital for combating infections; gene therapy restored immunity by supplying their lymphocytes with a working copy of the ADA gene.

Last December three teams of gene therapy researchers received permission to tackle one of the most common lethal hereditary diseases: cystic fibrosis (CF), which affects one in every 2,500 white babies. (The disease is much rarer in other races.)

The healthy version of the so-called CF gene produces a protein necessary for making normal mucus in the lungs. Most people have either one or two healthy copies of the gene, but people with CF have two abnormal copies, leaving them unable to produce normal mucus. The main result of this incurable condition is thick, sticky mucus in the lungs that builds up and encourages severe respiratory infections that usually prove fatal by age 30.

All three research teams are trying the same basic strategy against CF, one that has worked well in animal studies: Package healthy copies of the CF gene into the genetic material of cold viruses, and then squirt the virus into the patients' airways. (As a safety measure, the viruses have been altered so they can't reproduce and harm the patient's cells once they're inside.)

If all goes as planned, the cold viruses will infect about 10% of the respiratory cells, inserting viral DNA into the cells and thereby furnishing them with the healthy CF gene as well. By spawning healthy copies of the CF protein, the newly implanted CF genes should help produce normal mucus that won't invite infections. [Editor's note: The first such trial experiment using a human subject was carried out at the NIH on April 17, 1993.]

Gene therapy clearly offers hope against the more than 4,000 inherited diseases—including CF and ADA deficiency as well as hemophilia and muscular dystrophy—that are known to result from a defect in single genes. But its potential uses are broader still. Genetic defects are known or suspected to play a role in major causes of death and disability, including cancer, heart disease, Alzheimer's disease and diabetes. Now, just over two years since gene therapy was first tried, some experts predict it could provide long-sought cures for these and other diseases.

The three established cancer therapies—surgery, chemotherapy and radiation—cure only about half of all people who develop cancer. One reason cancers defeat their human hosts is that tumors are usually "invisible" to the immune system. While immune cells will attack invading bacteria and viruses, they often treat cancer cells as if they belong in the body. In attacking cancer through gene therapy, several research teams are trying to crank up the immune response in one of two ways: revving up immune cells or making cancer cells "visible" to the immune system so it will recognize and attack them.

In the first gene therapy cancer trial ever, ongoing since early 1991, researchers led by Dr. Steven A. Rosenberg, chief of surgery at the National Cancer Institute, will eventually treat up to 50 severely ill patients who have melanoma. The researchers first collect white cells found within a patient's tumor, since *these* immune cells have shown they can home in on cancerous tissue. Using a genetically altered virus, the researchers then "infect" these cells with a gene that produces a powerful cancer-fighting protein called tumor necrosis factor. Thus fortified, the immune cells are grown in large quantities and then reinfused back into the patients in gradually increasing doses.

In a more direct approach to treating melanoma, researchers at the University of Michigan in Ann Arbor led by Dr. Gary Nabel, an associate professor of internal medicine and biological chemistry, have begun injecting a gene directly into the tumors of 12 patients—in effect using DNA as an experimental drug or cancer "vaccine." Rather than using viruses to ferry the gene into tumor cells, the Michigan researchers encase millions of copies of it in tiny fat globules that slip through the membranes of the cells and deposit the gene within.

Once inside tumor cells, the gene should prompt the cells to produce a special protein—an antigen—that prompts immune cells to attack. The protein will protrude from the surface of the tumor cells, betraying them to the immune system in much the same way that a hoisted periscope can jeopardize a submarine. Studies in mice suggest that once the immune system is tipped off to the cancer's presence, it will also track down and kill tumor cells not modified by the new genes.

Last December, in an ingenious attempt to expose tumors to attack, researchers began treating brain tumor patients by introducing a gene from the herpes simplex virus into their tumor cells—essentially making them resemble a herpes virus. Ideally, the tumor will be eliminated when treated with the antiherpes drug ganciclovir.

A second group of experiments aims to cure cancer by another route: correcting the genetic defects that cause it.

Research has shown that cancer is a disease with a definite genetic basis. Even when environmental chemicals clearly play a role—as in the case of cigarette smoke and lung cancer—the carcinogens do their damage by mutating genes that control cell division.

Genes known as proto-oncogenes regulate the orderly cell division that occurs continuously throughout the body. But a mutation can turn a proto-oncogene into an oncogene, or cancer gene, which triggers the uncontrolled cell division that results in tumors. A second class of genes, called suppressor genes, normally rein in oncogenes, but a mutation can make them useless.

Cancer appears to be the culmination of a series of such mutations that ultimately pushes cells to multiply out of control. Gene therapy could conceivably cure or even prevent cancer by replacing a defective regulatory gene with a normal gene. One experiment that has generated considerable excitement among scientists will attempt to do exactly that.

Last fall, Dr. Jack Roth, chairman of thoracic surgery at the M.D. Anderson Cancer Center at the University of Texas in Houston, received permission to try gene therapy on 14 patients extremely ill with lung cancer, the deadliest cancer in the U.S. Roth will first surgically remove as much of a patient's tumor mass as possible and then inject the remaining tumors with viruses genetically engineered to contain one of two foreign genes. One gene supplants flawed copies of the tumor cells' suppressor gene; the second targets the cells' out-of-control oncogene.

That second gene is a synthetic strand of DNA specially created to mirror the oncogene in tumor cells—a strategy known as "antisense." Ideally, the antisense gene will latch onto the oncogene, gumming it up so it can no longer produce its dangerous proteins.

Animal experiments are now under way that could soon lead to gene therapy for several other major disorders. Among them:

• **Heart Disease.** More Americans die from heart disease each year than from any other condition. One research goal is to spawn new blood vessels in heart muscle deprived of blood due to a heart attack. Researchers at the University of Chicago School of Medicine are using genes that produce growth factors to try to grow new vessels in the damaged heart muscles of animals.

A major problem with the two main operations for opening up diseased arteries—coronary bypass and balloon angioplasty—is that often they must be repeated because proliferating muscle cells in the artery walls soon block the vessels again. The genes involved in muscle-cell growth were recently identified, and in separate experiments the Chicago researchers hope to block their action through use of the antisense strategy described earlier.

• **Hemophilia.** People with this hereditary disorder have defects in genes that produce clotting factors, needed to halt bleeding. The genes for several of these clotting factors have been identified, and several teams of researchers are developing techniques to insert these genes into human tissue.

• **Parkinson's disease.** The shaking, rigid posture and unbalanced walk that occur in Parkinson's disease patients are caused by lack of the chemical dopamine, which is needed to transmit impulses among nerves in the brain. Inserting the gene responsible for making dopamine into brain cells might correct the problem; researchers from several U.S. medical schools have worked together to devise a way to do that. Using a modified herpes simplex virus to deliver the key enzyme for producing dopamine, they've observed significant improvement in rats with a condition resembling Parkinson's disease. At a symposium last fall, the researchers described their results as "preliminary but encouraging."

• **Alzheimer's disease.** While the cause of Alzheimer's disease is unknown, the absence of some chemical could play a role, and one possible candidate is nerve growth factor (NGF). Using the modified herpes simplex virus as a vehicle, researchers have successfully introduced the NGF gene into rat nerves, where it produced new NGF. If missing NGF is in fact part of the cause of Alzheimer's, such experiments could pave the way for its eventual treatment.

While gene therapy certainly offers hope against some incurable and rare hereditary disorders, it's premature to consider the treatment a panacea. The experiments described here—especially those involving laboratory animals—are still only experiments, and results are just starting to trickle in. Yet even at this early stage it seems clear that gene therapy will play an important role in solving some of the most intractable of human diseases.

"Someday people will look back on the era before gene therapy in the same way we look back on the era before antibiotics and vaccines," predicts Dr. Rochelle Hirschhorn, a professor of medicine at New York University School of Medicine and a genetics researcher. "It is now possible to think about treating a whole series of diseases with a one-shot therapy that would last a lifetime."

GENETIC ATTACKS ON AIDS READIED[7]

The AIDS virus continues to elude all conventional attacks. Find an effective drug and the virus quickly mutates to sidestep the obstacle. Make a vaccine and the virus changes its coat to become invisible to the immune system's antibodies.

In search of fresh approaches, some molecular biologists say it is a mistake to treat AIDS as if it were caused by an ordinary virus like those of influenza or polio. Instead, they say, AIDS should be seen as a disease of DNA. Since H.I.V., the virus that causes AIDS, integrates itself into the DNA of the chromosomes, they argue, the methods of gene therapy and molecular biology should be used to attack the virus as it subverts the cell's genetic machinery, almost as if it were a mutant human gene.

The perspective of AIDS as a disease of DNA has helped inspire several new and ingenious approaches. Although these may prove no more successful than the conventional efforts, they represent the most sophisticated tools that molecular biologists now command. Two such schemes are now ready to be tried in people with AIDS and others will be tested in patients later this year. As always in clinical studies, these initial trials are intended to establish safe modes of treatment, not to accomplish a cure at this stage.

[7]Reprint of article by Gina Kolata from *The New York Times* pC1 My 31 '94.

The challenge of a gene therapy approach to AIDS is to choose the right viral genes to sabotage. 'You can target different proteins that are controlling different parts of the viral life cycle,' said Dr. Gary J. Nabel, a Howard Hughes researcher at the University of Michigan Medical Center in Ann Arbor. 'You can target proteins that are needed for the virus to get into the cell, to get into the nucleus, to get out of the nucleus.'

Dr. Nabel and his colleagues are aiming at a small but critical cog in the machinery by which H.I.V. assembles new viruses. It is a protein made by an H.I.V. gene called rev and it has a humble but essential duty, that of transporting copies of most of the virus's genetic messages from the nucleus of a cell into the surrounding cytoplasm.

Without the rev protein, the genetic messages needed to make new viruses stay bottled up in the cell nucleus and are unable to direct the synthesis of more viruses.

Dr. Nabel's idea is to add a gene to cells that makes a subtly altered version of the rev protein, one that readily latches onto the genetic messages created by the virus, but that is unable to transport the message out of the nucleus.

Dr. Nabel said that this gene therapy worked well in human cells growing in the laboratory. He is about to start a study testing it with 12 people with H.I.V. infections. About a billion white blood cells, a thousandth of the body's total, will be taken from each patient and treated with a vector that inserts the variant rev gene into the chromosomes of each cell.

Dr. Nabel is testing two different vectors. One is a virus that inserts the rev gene into the cell's DNA. The other is a gold bullet. The spaghetti-like DNA is wrapped around a tiny gold pellet and the pellet is shot into the cells. The cells take up the rev gene and spit out the gold, Dr. Nabel said. The treated cells will then be reinjected and the patients monitored to follow their progress. If the variant rev gene does protect his patients' cells, Dr. Nabel and his colleagues will then treat a larger proportion of the body's white cells, in an attempt to treat the disease.

A Second Strategy

Dr. Roger A. Pomerantz, a molecular virologist at Thomas Jefferson University in Philadelphia, is also targeting the rev gene but has found a different way to block it. His method is to inactivate the protein with a specially designed antibody. Ordinarily, antibodies are large molecules that are excreted by cells that produce them. They float free in the bloodstream as part of the body's surveillance system, attaching themselves to foreign molecules.

Dr. Pomerantz has designed an antibody that recognizes the virus's rev protein. He inserts the gene for the antibody into white blood cells by putting it in a virus. Since the antibody stays inside the white blood cell where it is made, it lacks the tags that signal the cell to export it. In the current issue of the *Proceedings of the National Academy of Sciences,* Dr. Pomerantz reports that his anti-rev antibody stops infected cells from producing H.I.V. He expects to start preliminary studies in people with AIDS within the next year.

In an added twist to these strategies of blocking viral genes with gene therapy, several groups of researchers have found ways to be sure that the added genes are quiescent unless and until they are needed. To do this, they attached an H.I.V. signal to the genes. This is the signal the H.I.V. uses to force a cell to start expressing its genes. So every time the cell starts to make H.I.V., it will make proteins that block H.I.V. from being made. Dr. Nabel, for example, said he added H.I.V. control regions to the genes that he would add with the gold bullets.

A third gene therapy approach is to concoct a gene that produces a powerful weapon known as a ribozyme. Ribozymes are strands of RNA that can function as enzymes, and they can be designed to attack other strands of RNA at specific chemical sites. This makes it possible to develop a ribozyme that will chew up RNA from H.I.V. without attacking any of the cell's own RNA molecules.

Several groups of researchers are pursuing the ribozyme strategy, including Dr. Flossie Wong-Staal of the University of California at San Diego. She has designed a ribozyme that recognizes a particular sequence of H.I.V., along with a method to insert a gene for the ribozyme into white blood cells. She and her colleagues are awaiting approval from the Food and Drug Administration to test the technique in patients.

If any of these gene therapies seems to discourage the virus, researchers say, the next step would be to combine the various gene therapies to narrow the virus's chance of escape.

Gene therapy, they say, presents a tougher challenge to the AIDS virus than the drugs and vaccines that it can avoid by mutating. The virus would have to change the shape of its rev protein completely to escape the anti-rev protein, while small mutations allow it to escape from the drugs and vaccines now available. Dr. Nabel said he had not seen any evidence that the virus could do that. This kind of consideration, molecular biologists said, gives them hope.

"An Intellectual Watershed"

"If you compare it to traditional drug development, in many ways it's an intellectual watershed," Dr. Nabel said. "Rather than pursuing drugs by trial and error, this allows us to use our knowledge base to design strategies."

Dr. John Rossi, a molecular biologist at the City of Hope Medical Center in Duarte, Calif., who is developing one such approach, said of gene therapy, "I think it is absolutely certain that it's going to make a difference."

But Dr. Rossi and others said that many hurdles remained for researchers and that a successful treatment, if it emerged, was probably years away.

Although Dr. Pomerantz said he, too, was optimistic, he added, "We have a long way to go."

"Clearly, this will be a row with more failures than successes," he said, "Don't expect next week that someone will yell, 'Eureka!'"

II. ETHICAL, SOCIAL, AND LEGAL ISSUES RAISED BY ADVANCES IN MEDICAL GENETICS

EDITOR'S INTRODUCTION

Medical science's success in isolating disease genes has meant that individuals can find out, through a simple test, if they are at risk for developing a given inherited disease. The pros and cons of getting tested, as well as the practical problems associated with screening large numbers of individuals, are discussed in the first article, written by Leslie Roberts and reprinted from *Science*. While genetic testing has many benefits—it can bring relief to individuals who test negative for an inherited disease and enable others to make informed decisions about childbearing, for instance—there are fears that genetic information could lead to discrimination in the workplace and denial of coverage by insurance companies. These issues are covered in the next two articles, "Use of Genetic Testing by Employers" and "Insurance for the Insurers: The Use of Genetic Tests." The ways in which an individual's right to privacy might be preserved is discussed in the fourth piece, from the *Journal of the American Medical Association*.

Another source of debate is whether or not gene therapy should be used not only to treat disease but also to "enhance" cosmetically an individual's features. This question is the subject of the fifth article, "Uses and Abuses of Human Gene Transfer." This section concludes with "DNA Goes to Court" by Robert M. Cook-Deegan, an examination of the reliability of DNA fingerprinting, which is increasingly being relied upon by the court system as a method of exonerating a defendant or helping to confirm his or her culpability.

TO TEST OR NOT TO TEST?[1]

For 5 years molecular geneticists predicted that they would bag the cystic fibrosis gene at any moment. But when Lap-Chee Tsui and Francis Collins finally did it last August, the medical and scientific communities were totally unprepared for what comes next—potentially the most widespread testing to date for carriers of a lethal genetic disease.

Within days of Tsui's and Collins's announcement, companies like Collaborative Research and Integrated Genetics began working on a test to determine whether individuals carry the cystic fibrosis gene and might thus be at risk of having a child with the disease. Cystic fibrosis is the most common lethal genetic disease of young Americans, and with 1 in 25 Caucasians a carrier of the disorder, the potential market is enormous. Geneticists are talking about screening the entire U.S. population, or at least all those of reproductive age, perhaps 100 or 200 million people.

Carrier screening has simply never been done on this scale before. The successful screening program for Tay-Sachs disease, for example, targeted only the 1 or 2 million Ashkenazi Jews of reproductive age.

But no sooner had the companies announced in November that their tests were ready than the American Society for Human genetics called for a voluntary moratorium on widespread population screening.

The immediate issue is that this new test is not yet definitive enough for mass screening. The gene defect that Tsui and Collins found accounts for many, but not all, cases of cystic fibrosis; the rest are caused by different mutations, perhaps ten of them, in the same gene. The current test detects only the known gene defect, which occurs in about 70% of all cystic fibrosis carriers. As a result, it would pick up just about half the couples at risk of having a child with this incurable disease. So even with a negative test, couples could still have an affected child.

In an unusual collaborative effort, nearly 50 labs worldwide are working flat out, searching for the other mutations that cause cystic fibrosis. Tsui, of the Hospital for Sick Children in Toronto,

[1]Reprint of article by Leslie Roberts from *Science* v 247 pp17–9 Ja 5 '90.

expects to find the majority within a year or so. And once they are found, the test should approach 100% accuracy.

But once a more definitive test is ready, there will still be major questions about how to screen that many people and do it well. Who, for example, should be screened, and when? Just Caucasians, in whom cystic fibrosis is most common, or blacks as well? And at what age? And who will educate the public about the test beforehand and, more importantly, explain what it means to all those who test positive?

The President's Commission for the Study of Ethical Problems in Medicine raised just these questions about cystic fibrosis screening in 1983 and urged policymakers to start planning for what was sure to come. But the questions are no closer to resolution now than they were then.

"We all are aware of how poorly sickle-cell screening was done in many areas of the country," says Collins of the Howard Hughes Medical Institute at the University of Michigan. "We don't want to repeat that travesty."

"The onus is on the screeners, before they unleash this technology, to show how to do it with the least negative impact and the most positive," says Michael Kaback, head of pediatrics at the University of California at San Diego and one of the chief architects, along with Arthur Beaudet of Baylor College of Medicine, of the genetics society statement. "But people get caught up in this simplistic view that the test will prevent disease and get on a fast track to deliver it."

Now the medical community is scurrying to get things in order while there is still time. There is a sense among most of those concerned, that this test is a big one and it behooves them to get it right, to set a precedent for other tests sure to follow.

But it is not clear how long testing will wait. "We've known for years that this day was coming and have done virtually nothing to prepare for it. Now there is a mad scramble—how do we control the test, or withhold it, while we sort out the issues?" says Keith Brown, president of Gene Screen, a Dallas-based genetic testing company. "But the train may have already have left the station."

For the time being, however, no one is quarreling with the genetics society's position that the existing test, if widely used, would do more harm than good. But the society does support its use for couples at high risk—those with a family history of cystic fibrosis. For them, says Beaudet, the benefits are clear.

If both parties carry the defective gene, there is a 1 in 4 risk with each pregnancy that the child will be born with cystic fibrosis, which is an autosomal recessive disease. That means that two copies of the gene must be passed on, one from each parent, for the child to inherit the disease.

And with the new test, there is a 50% chance that the couples at risk would be identified and could be helped. If the woman is already pregnant, the couple could go ahead with prenatal diagnosis to determine whether the child is affected, explains Beaudet. If the woman is not pregnant, other options exist—adoption, artificial insemination, or monitoring future pregnancies with prenatal diagnosis and with the possibility of having an abortion, if they so choose. Until now, carrier testing has been available only to those couples with a living child with the disease—a small fraction of those who harbor the gene.

But what's far more likely statistically than finding two carriers, says Beaudet, is finding that one person tests positive and the other negative. And that gets tricky because the person with the negative test could still be a carrier of one of the as yet unidentified mutations. Before testing, the couple would have had a 1 in 2500 risk of having a child with cystic fibrosis, which is the risk for the general population. After testing, their risk would have jumped to about 1 in 400. And there is nothing anyone can do to resolve their uncertainty. For every couple who could be helped by the test, there will be about 25 more in limbo, in which just one is a carrier. Says Beaudet: "My concern is that it will get a lot of people worried with no good way to resolve it."

This is not to say that couples who are not at high risk shouldn't be tested anyway, as long as they are informed of all the uncertainties. Says Beaudet: "This is not a ban on carrier testing." Collins agrees. "I don't think anyone should deny a couple who wants testing. But I don't think we as a profession should push it."

At this point, the companies offering the test seem to agree. Even before the genetics society released its statement in November, Integrated Genetics, in Framingham, Massachusetts, had decided to offer its test only to those with a family history of the disease, says general manager Peter Lanciano. When a physician requests the test, says Lanciano, he must vouch for the family's history. In the interim, the company is promoting the test only to academic genetic centers and private genetic practices—"the only ones who can understand the test and provide support services," says Lanciano.

Other companies, like Gene Screen and Collaborative Research of Bedford, Massachusetts, while not refusing the test to anyone, say they are at least not pushing it. "Our position is that we will accept specimens from any physician who requests the test," says Brown of Gene Screen. "But we are not promoting it to OB/GYNS. Believe me, they are not ready for it," he says, citing a survey the company recently conducted on physician awareness about the disease.

It is not all altruism, Brown is the first to admit. "CF presents an opportunity to us, but it has to last. We can't get off to a false start or have it blow up in our face."

Meanwhile, couples are already requesting the test, says Collins, and despite the society's statement, which was intended to assure them otherwise, some physicians are worried that they will be hit with a malpractice suit if they don't offer it.

"We all feel public pressure to get on with it," says Nancy Lamontagne of the National Institute of Diabetes and Digestive and Kidney Diseases. And that is why she, along with Elke Jordan of the NIH genome center, and the American Society of Human Genetics are hastily putting together a February workshop to tackle the plethora of questions surrounding the new test.

One of the first questions is simply how sensitive the test must be for widespread screening. Most agree that 70% is not good enough, but is 90%? 95%? 99%? The same question came up with alpha-fetoprotein testing, which detects neural tube defects and other problems, says Neil Holtzman of the Johns Hopkins University School of Medicine. "The question then was how can you withhold tests and continue to see kids born with these defects?"

Even fundamental questions like who should be tested, and when, must be sorted out. Although cystic fibrosis is primarily a disease of Caucasians, the gene does occur, at about one-tenth the frequency, in American blacks as well. Should everyone be screened, as Kaback advocates, or should screening be limited to Caucasians, as the President's ethics commission concluded in 1983? And should they be tested as newborns, adolescents, or later in life?

The goal is to test people before they conceive, while they still have a number of options. But how do you reach them beforehand? It hasn't worked well in the past, concedes Beaudet. "If you look at what has happened with other diseases, most couples are tested when they are pregnant."

Where the test is offered will make a difference. The most efficient way to reach people of reproductive age is to offer the test as part of obstetrical care, says Holtzman, perhaps piggybacking it onto prenatal tests already offered. If a woman tests positive, then her partner would come in for testing.

But if the goal is to provide information to make informed reproductive decisions, then an obstetrician's office may not be the way to go, counters Kaback. If screening is offered through a doctor's office, he says, it will almost invariably be done when the woman is pregnant.

The alternative would be a community-based screening program, perhaps modeled on the Tay-Sachs program that Kaback helped start in the early 1970s, which offered testing through synagogues, community centers, and the like. . . But for cystic fibrosis, the numbers are daunting. "There might be a way to figure out how to do it logistically," says Elena Nightingale of the Carnegie Corporation of New York, "but where do you find the workers when you are talking about screening that many people?"

And who is going to educate the public about the test and then counsel those who are positive? "Screening without education and counseling would be a catastrophe," asserts Kaback.

Collins agrees: "One in 25 Caucasians is a carrier. That means 8 million Americans. And every one of them deserves an explanation. The problem, says Jessica Davis, codirector of the Center of Human Genetics at Cornell University Medical College, is that "there simply aren't that many card-carrying clinical geneticists and counselors around."

All this assumes that everyone will want to be tested, which is not at all clear. Alpha-fetoprotein testing, for example, is now routinely offered to pregnant women in this country who receive prenatal care, but only about 40% elect to have it, says Lanciano of Integrated Genetics. He believes that, likewise, demand for the cystic fibrosis test may be much lower than many geneticists are now predicting. So does Holtzman, who points out that just one-fourth of young Jewish adults are screened for Tay-Sachs disease. And mass screening depends on a reliable and cheap test, which so far does not exist. The test is now going for anywhere from $125 to $225 a pop; for mass screening, says Brown, the upper limit is about $50.

Kaback is calling for pilot programs, similar to those he ran for Tay-Sachs, to evaluate, among other things, which educational approaches work best, how many people elect to be tested, just what the counseling needs are, and "how much fear we create."

And even before those studies are done, he and others say, some type of centralized quality control must be set up to monitor the laboratories already offering the test. They'd better hurry, says Brown of Gene Screen. "To [expect us to] wait until we get 99% of the mutations and a national program is defined in 2½ years, that's kind of dreaming. The genetics community is thinking about how to make it happen ideally. Forget it, that game is already lost. The question is, how can the genetics community make it to happen better?"

It's just a matter of time, Brown says, before *Cosmo* or *Redbook* runs an article that will educate a lot of women about the test. "It will educate lawyers too. And the first lawsuit against someone who didn't offer the test will get a lot of attention." At some point he says, one of the companies is going to decide that the test is good enough, that obstetricians are ready, and "go for it." Adds Brown: "And once one company starts offering it, it will be very difficult for others to hold back."

USE OF GENETIC TESTING BY EMPLOYERS[2]

Over the next 15 years, under the auspices of the federal government's "human genome project," scientists will try to map in detail each of the human's cell's estimated 100,000 genes. The knowledge derived from the project will enable physicians to detect an increasing number of diseases and predispositions for disease. It is expected that researchers will identify genes that contribute to the development of Alzheimer's disease, alcoholism, coronary artery disease, the different forms of cancer, and virtually every other illness. In addition to enhancing the ability of physicians to diagnose disease, the knowledge from the genome project is expected to result in better preventive and therapeutic measures.

Potential applications of information gained from the human

[2]Reprint of article by the Council on Ethical and Judicial Affairs, American Medical Association, from the *Journal of the American Medical Association* v 266 pp1827–30 O 2 '91.

genome project extend well beyond the setting of medical care. Employers, insurers, and law enforcement agencies all will have uses for genetic-testing techniques. In many cases, these uses will provide important social benefits. DNA fingerprinting can establish with greater certainty the identity of a criminal; it can also exonerate the innocent defendant. However, our experiences with genetic and other medical testing suggest that abuses may occur. Some companies may have restricted employment opportunities for individuals who carry the sickle cell trait, even though no scientific basis for the restrictions existed. In addition, employment discrimination has occurred repeatedly against individuals because of their medical problems. Previously, irrational fears led employers to deny jobs to patients with cancer or epilepsy. Individuals infected with the human immunodeficiency virus continue to be victims of employment discrimination.

In this report, the council will address the use of genetic testing by employers to identify employees (or potential employees) at risk for developing certain diseases. Genetic testing by employers will involve the participation of physicians. This report will propose guidelines to help physicians assess when their participation in genetic testing by employers is appropriate and does not result in unwarranted discrimination against individuals with genetic abnormalities.

Workplace Testing

Employers will have a number of potential justifications for genetic testing in the workplace. In some cases, there may be an argument in favor of testing for public health reasons. Companies have expressed concern about the possibility of an employee's genetic susceptibility to illness from exposure to a chemical or other substance in the workplace. In addition, employers may not want to hire individuals with certain genetic risks for jobs that bear on the public's safety. Other justifications are based not on concerns about health but on concerns about costs, specifically the costs to the *company* of hiring workers with a genetic risk of disease. Individuals who have a heightened risk for certain illnesses may be less attractive as employees; on average, they may be able to spend fewer years in the work force, and they may impose greater health care costs on the employer.

Since the acceptability of genetic testing depends on the purposes for which the testing is proposed, each of the proposed justifications for testing will be considered separately.

FUTURE UNEMPLOYABILITY

Employers may be reluctant to hire individuals who have a genetic predisposition for developing a disabling illness, such as cancer or coronary artery disease, because these individuals may become prematurely unable to work. By excluding those at risk, employers may be able to lower their costs of recruitment and training.

As an ethical matter, however, future unemployability is not an adequate basis for performing genetic tests. Genetic tests are poor predictors of disease and even poorer predictors of disabling disease. Genes are often characterized by incomplete penetrance; that is, many individuals who carry the gene will never show manifestations of the gene. When the gene manifests itself, it will be characterized by variable expression—the extent of the gene's effects may differ widely from person to person. Among individuals with sickle cell anemia, some die within the first years of life, while others survive into their 50s. In many cases, behavioral modification can limit gene expression. Patients at risk for diabetes can modify their diet, as can patients at risk for coronary artery disease. Even in cases in which the gene will ultimately cause disabling disease, the effects of the gene may not appear for some time. For example, the onset of Huntington's disease does not occur until the patient is between the ages of 30 and 50 years. Consequently, the use of genetic tests would result in individuals' being denied employment well before they became unable to work. In sum, genetic tests would have a high false-positive rate and, therefore, would result in many individuals' being denied employment unfairly.

Exclusion on the basis of future unemployability is problematic also because it can seriously undermine the principle that underlies the protection of disabled individuals from employment discrimination: disabled individuals should not be denied employment when their disability does not interfere with their ability to perform a job. If individuals could be denied jobs because of future inability to perform, then anyone infected with the human immunodeficiency virus could be denied employment, as could individuals with other diseases that lead to premature loss of the ability to function in the workplace.

Legal rules are in accordance with this ethical analysis. When construing statutes that forbid employment discrimination against the disabled, courts have consistently rejected employers' arguments that they should be able to deny employment to applicants whose future work might be compromised by health problems. Although these statutes do not apply to a large percentage of the work force, the recently enacted Americans With Disabilities Act will prohibit virtually all employers with 15 or more employees from discriminating on the basis of disability. Under the Disabilities Act, medical testing will not be allowed unless the testing is related to the applicant's actual ability to perform the job.

Employers should not be forbidden entirely from considering a person's future qualifications for a job, and the Disabilities Act will probably permit some consideration of future employability. An individual who would not be able to continue employment for more than a very short time, whether for health or other reasons, need not be treated the same as someone who can work on a long-term basis. Important factors in assessing future employability are the likelihood that the employee would no longer be able to work and the length of time before the ability to work would be lost. However, employers can make valid distinctions on the basis of an individual's health status without using genetic tests. People who will become unable to work in a short time can be identified by medical testing that measures the effects of genes rather than the genes themselves.

Increased Health Care Costs

Employers may not want to hire individuals with a predisposition for cancer, Alzheimer's disease, or other illnesses since these individuals might impose higher health care costs on the employer.

Many of the considerations that counsel against genetic testing to assess future employability apply here as well. Because of incomplete penetrance, variable expression, and delayed manifestation, genetic tests have poor predictive value also when used as a method for limiting health care costs. In addition, protection against discrimination on the basis of disability would be vitiated if health care costs could be used as a criterion of employment. Individuals with disabilities typically have higher-than-average health care bills. Consequently, the Americans With Disabilities Act does not recognize higher health care costs as a basis for screening potential employees. The Disabilities Act does permit employers to take health risks into account when issuing employee health and other insurance. However, it expressly prohibits employers from using risk underwriting for insurance as a subterfuge to evade the antidiscrimination purposes of the act.

From the perspective of society's economic interests, denying employment on the basis of higher health care costs rarely makes sense. Whether or not the person with a genetic abnormality is employed, society will face the same health care costs. (An important exception would occur when a person has a genetic susceptibility to injury from exposure to a chemical in the workplace, a situation that is discussed under the "Susceptibility to Workplace Exposures" section.) If the person is denied employment, however, there can be no countervailing benefit from the person's ability to contribute productivity in the work force. Thus, if the disabled are working, society's economic interests are better served, as are principles of equity and justice.

There also would not be any unfairness to employers. Since all employers would have the same ethical obligations, any increases in costs would apply across the board. However, if some employers end up with a disproportionate burden of health care costs, it would be appropriate for the government to assist in the form of high-risk insurance pools, tax credits, or other subsidies.

Public Safety

In some cases, employers may want to use genetic testing to protect the public's safety. For example, employers of physicians or airline pilots may want to test for the gene that contributes to the development of Alzheimer's disease when such a test exists.

Although ensuring the health of employees whose work bears on public safety is an important responsibility of employers, genetic testing is not an appropriate tool for meeting that responsibility. As when used for other purposes, genetic tests will have poor predictive value when used to identify workers who might pose risks to public safety. Incomplete penetrance, variable expression, and delayed manifestation are problems here too. Thus, most individuals who would be excluded from employment by genetic testing never would have presented a heightened safety risk. Antidiscrimination law has recognized that individuals might be wrongly denied jobs because of speculative safety risks. Consequently, before a person can be excluded, an employer must show that there is a significant or reasonable likelihood of harm to others from having a person with a genetic risk of disease employed in the workplace.

Genetic tests are not only generally inaccurate when used for public safety purposes, but also unnecessary. A more effective approach to protecting the public's safety would be routine testing of a worker's actual capacity to function in a job that is safety-sensitive. Airline pilots, for example, currently undergo physical examinations every 6 months. Companies that employ bus drivers or ship operators have begun to use simple neurobehavioral testing on a frequent basis to test for impairment by drugs and other causes. Routine and regular functional testing could be used to detect those individuals who become incapacitated by a genetic disease as the disease manifests itself. In addition, the testing would detect those whose incapacity would not be detected by genetic tests, either because of a false-negative test result or because the incapacity was caused by something other than the disease being tested for.

Functional testing might also be required by the Disabilities Act. According to the act, an individual cannot be excluded from the workplace on grounds of safety if "reasonable accommodations" by the employer would eliminate the safety risk. If functional testing would be more precise than genetic testing at identifying workers who pose a safety risk, then functional testing would likely be viewed as a reasonable accommodation for the employer.

Susceptibility to Workplace Exposures

Since at least the 1960s, there has been interest in screening workers for genetic susceptibility to injury from chemicals or other substances in the workplace. Some occupational health experts have argued that genetic tests can be used to identify workers who are particularly at risk for injury from workplace toxins. In fact, black employees have been screened for the presence of sickle cell trait because of concern that exposure to nitro or amino compounds would result in sickling of the blood cells. Male workers have been screened for the sex-linked genetic abnormality of glucose-6-phosphate dehydrogenase deficiency because of concern that exposure to oxidizing chemicals would precipitate hemolytic anemia. Genetic screening has also been conducted to identify workers with $alpha_1$-antitrypsin deficiency on the ground that respiratory irritants might cause chronic obstructive lung disease.

Although these genetic tests have been used for research and to advise workers of potential risks, they also may have been used inappropriately to exclude affected individuals from the workplace. For instance, the apparent exclusion of workers with sickle cell trait was based on theoretical considerations that had no basis in fact. To date, there is insufficient evidence to justify the use of any existing test for genetic susceptibility as a basis for employment decisions.

With greater understanding of genetic disease, researchers may develop tests that are more useful in identifying individuals at genetic risk for occupational injury. However, it is doubtful that an abnormal result on one of those tests will be sufficient justification to deny employment to the affected person. The poor predictive value of genetic tests is relevant in this context as well. Many individuals with abnormal test results will never express the gene, will express the gene mildly, or will not express it for a long time. Consequently, many people would be denied employment unfairly.

There is also a serious concern with false-negative test results. If companies adopt a policy of excluding hypersusceptible individuals from the workplace, they may relax their efforts to eliminate potential toxins from the workplace, since the remaining workers will be able to tolerate a higher level of exposure to the toxins. However, because of testing inaccuracies, some affected individuals will not be detected. If they are hired, they will then face an especially elevated risk of injury.

Protecting workers from occupational injury can be achieved much more effectively by offering workers the opportunity to be monitored both for their exposure to potential toxins and for adverse health effects from the toxins. When employees are exposed to lead, levels of lead in the workplace are regularly measured to prevent excessive exposures. Similarly, in workplaces where radioactivity is present, the amount of radioactivity that workers are exposed to is routinely monitored to ensure that the employees do not receive an inordinate dose. If workers develop too great an exposure to the toxin, they should be transferred to a safer job without loss of salary, benefits, seniority, or opportunity for advancement. Under occupational safety and health law, when employees develop excessive blood lead levels from workplace exposure, the employer must provide an alternative job with existing pay and benefits for up to 18 months.

In addition to offering routine monitoring, employers should notify applicants for employment of the occupational risks that they would face in the job and inform them that genetic susceptibilities might increase their risk. The applicants could then have their own physicians perform genetic testing and could decide whether they are willing to assume the risk of exposure if the test results are abnormal. Although there is insufficient justification for employers to exclude workers with genetic susceptibilities to injury, potential employees should be able to decide that they are unwilling to accept even a small risk of injury.

It is conceivable that informing applicants and monitoring workers might not be adequate precautions. For instance, although there are no current examples, researchers may discover that a disease develops so rapidly that significant and irreversible injury would occur before monitoring could be effective in preventing the harm. In such a case, testing may have a role in identifying those who are genetically susceptible to the disease, so that they can be excluded from the workplace.

However, before genetic testing could be used for exclusionary purposes, other requirements would have to be met. The employer would have to demonstrate that the genetic tests are highly accurate, with sufficient sensitivity and specificity to minimize the risk of false-negative or false-positive results. In addition, empirical data would have to demonstrate that the genetic abnormality results in an unusually elevated susceptibility to occupational injury.

Employers would also have to show that it would be too costly to reduce the risk to the susceptible worker by lowering the level of the toxic substance in the workplace. To demonstrate undue cost, the employer would have to show that the costs of improving the safety at the workplace are extraordinary, relative to other costs of production. Since the alternative to cleaning up the workplace is genetic testing and exclusion, the employer would also have to show that the costs of improving the safety of the workplace are extraordinary, relative to the costs of testing potential employees for genetic susceptibility. These requirements would ensure that the costs of using the toxic substance are not placed on a few individuals but on society as a whole. Since society as a whole benefits from the use of the toxin, society as a whole should pay for its use.

Finally, genetic testing should not be performed without the informed consent of the employee or applicant for employment.

Under the Americans With Disabilities Act, if employers use genetic testing to exclude current workers from certain jobs, they may be obligated to offer them alternative employment. Since genetic testing is ethically permissible only under very limited circumstances, the obligation to provide alternative employment would likely apply to only a small number of workers. Although the obligation does not apply to applicants for employment, the Disabilities Act permits preemployment genetic tests only in limited circumstances. Although an employer may condition an offer of employment on the results of genetic testing that is job related, testing may not be performed until after a conditional offer of employment has been made.

Evading the Prohibitions on Testing

Even if employers do not use genetic testing, they may still be able to find out whether their workers have certain genetic predispositions for disease (M. A. Rothstein, JD, University of Houston Health Law and Policy Institute, written communication, March 1991). Employers will often have access to the medical records of their employees. In some cases, medical records are obtained if questions arise about the employee's ability to resume work after an illness or accident. Although the employer would not need to receive the part of the medical record that includes genetic information, unnecessary information is often disclosed in response to a request for medical records. The patient's genetic information may also be disclosed if the patient receives treatment related to a genetic condition and files a claim for health insurance benefits. Measures will have to be developed to protect the confidentiality of a patient's genetic status.

For the reasons described in this report, the Council on Ethical and Judicial Affairs has developed the following opinion:

Opinion 2.131: Genetic Testing by Employers

As a result of the human genome project, physicians will be able to identify a greater number of genetic risks of disease. Among the potential uses of the tests that detect these risks will be screening of potential workers by employers. Employers may want to exclude workers with certain genetic risks from the workplace because these workers may become disabled prematurely, impose higher health care costs, or pose a risk to public safety. In addition, exposure to certain substances in the workplace may increase the likelihood that a disease will develop in the worker with a genetic risk for the disease.

1. It would generally be inappropriate to exclude workers

with genetic risks of disease from the workplace because of their risk. Genetic tests alone do not have sufficient predictive value to be relied on as a basis for excluding workers. Consequently, use of the tests would result in unfair discrimination against individuals who have abnormal test results. In addition, there are other ways for employers to serve their legitimate interests. Tests of a worker's actual capacity to meet the demands of the job can be used to ensure future employability and protect the public's safety. Routine monitoring of a worker's exposure can be used to protect workers who have a genetic susceptibility to injury from a substance in the workplace. In addition, employees should be advised of the risks of injury to which they are being exposed.

2. There may be a very limited role for genetic testing in the exclusion from the workplace of workers who have a genetic susceptibility to occupational illness. At a minimum, several conditions would have to be met:

• The disease develops so rapidly that serious and irreversible illness would occur before monitoring of either the worker's exposure to the toxic substance or the worker's health status could be effective in preventing the harm.

• The genetic testing is highly accurate, with sufficient sensitivity and specificity to minimize the risk of false-negative and false-positive test results.

• Empirical data demonstrate that the genetic abnormality results in an unusually elevated susceptibility to occupational illness.

• It would require undue cost to protect susceptible employees by lowering the level of the toxic substance in the workplace. The costs of lowering the level of the substance must be extraordinary, relative to the employer's other costs of making the product for which the toxic substance is used. Since genetic testing with exclusion of susceptible employees is an alternative to cleaning up the workplace, the costs of lowering the level of the substance must also be extraordinary, relative to the costs of using genetic testing.

• Testing must not be performed without the informed consent of the employee or applicant for employment.

INSURANCE FOR THE INSURERS: THE USE OF GENETIC TESTS[3]

Should health and life insurance companies use genetic tests to determine applicants' eligibility or to set rates, either by direct screening or by asking applicants about an individual history of genetic testing? Concerns exist that individuals will be denied access to health care and that the government, left to foot the bill for those denied private coverage, will have an increasingly costly health care budget. The fear also has been raised that employers might screen employees to eliminate those who could cause a rise in group premiums, or that individuals might be less likely to volunteer for research knowing that information gained in that manner might later be used against them when applying for insurance.

Such concerns highlight the ambiguous role of insurance companies in our society. On the one hand, individuals, particularly sick individuals, desperately want to have access to or maintain their private insurance; at the same time, insurers argue that they are private businesses, that they have a responsibility to generate profits for their stockholders, and that it is no more their responsibility to assure access to care than any other private business.

Rapid advances in molecular biology and human genetics are yielding technology that, among its other applications, can be used by insurance companies to refine further their risk-screening procedures. Whether these procedures should be put to use at all, and, if so, how much screening will be considered acceptable, is a matter for public policy. In this essay I question whether the precedents set by the U.S. insurance industry are an adequate basis for policy of this kind. I begin by examining those precedents—that is, by exploring how the industry came to be what it is today.

History of the Insurance Industry

The first private health insurance plan in this country was established in 1929 to cover hospital expenses for 1,250 Texas

[3]Reprint of article by Nancy E. Kass from the *Hastings Center Report* 22, v 6 pp6–11 '92. Copyright © 1992 by The Hastings Center. Reprinted with permission.

schoolteachers. By the mid 1940s such "Blue Cross" hospital plans existed in forty-three states. The first plans designed to reimburse for physician services were instituted in 1939, and by the 1950s these came to be known as Blue Shield. From their inception until 1986, the "Blues" were nonprofit and tax-exempt enterprises. Initially, their premiums were based on "community rating," whereby all subscribers in a given geographic region were charged the same rates. The healthy insured thus subsidized the sick.

Commercial insurance companies emerged in the 1940s. To compete for employer contracts, they offered premiums based on the new concept of "experience rating": employers were charged based on the actual claims experiences of their own employees. Since workers tend to be healthier than the general population, experience-rated premiums were lower than community-rated ones. The nonprofit companies soon felt the effects of the shift and had little choice but to alter their own rating practices. In this manner a system was established in which those who were sicker, unemployed, or not part of a group were charged higher rates than lower-risk or group members.

Historically, there has been great variation in the degree to which the Blues served as the organizations of social welfare their tax-exempt status suggests. While some plans offered coverage to individuals who could not obtain it through a commercial carrier, others had underwriting practices comparable to the practices of commercial companies. A study conducted by the General Accounting Office in 1985 showed that 38 percent of the Blues offered an open enrollment period in which applicants could buy health insurance, regardless of their health conditions; none of the commercial companies did so. But due to a belief that the Blues' screening and rate-setting practices were not vastly dissimilar from those of the commercial industry, Blue Cross and Blue Shield plans lost their tax-exempt status in 1986.

A bill has been proposed that would restore tax-exempt status to plans that "make a contribution to social welfare." To be eligible, plans would need to offer a continuous open enrollment period, have at least 35 percent of their premiums community rated, have at least 10 percent of their policies covering individuals and small groups, and not have any stockholders.

Insurance Coverage Today

Insurable events are assumed to be random (that is, unpredictable), free of "moral hazard" (meaning that having insurance does not provide the insured with an incentive to make a claim, as would be the case with arson), and free of "adverse selection" (which is to say the insured and the insurer make decisions based on equal amounts of information).

In accordance with the risk classification begun in the 1940s, unfair trade practices acts exist in each of the fifty states and the District of Columbia. These acts prohibit unfair discrimination "between individuals of the same class and of essentially the same hazard in the amount of premium, policy fees, or rates charged for any policy or contract of health insurance." However, and more germane to the genetic testing issue, unfair trade practices acts have been interpreted to justify treating individuals at *different* risk differently. It is the insurance industry's perspective that an insurance company "has the responsibility to treat all its policy holders fairly by establishing premiums at a level consistent with the risk represented by each individual policyholder." Such is the concept of insurance "equity": premiums must correspond fairly to the differences in risk posed by individual policy holders.

Whereas applicants for group health insurance typically are not screened to determine their individual risks, individuals and small groups typically are. Applicants either must fill out a health history questionnaire, or, about 20 percent of the time, they must furnish a copy of their medical record or a statement from their physician. Applicants are classified as standard, substandard, or denied. Seventy-three percent of individual applicants for commercial policies and health maintenance organizations, and 83 percent of individual applicants for nonprofit plans are classified as standard. Those accepted as substandard (a classification that does not exist in an HMO system) are offered a policy that comes with an exclusion waiver, a higher premium, or both. Most insurers deny coverage altogether to applicants whose probability of disease exceeds three times the average for their age and sex.

Since the 1868 U.S. Supreme Court ruling in *Paul v. Virginia,* insurance regulation has been considered the domain of state rather than federal governments. This was upheld in another U.S. Supreme Court case in 1944 as well as in the 1945 McCarran-Ferguson Act, which legislated the continued regulation of the insurance industry by the states in all cases other than interstate transactions.

New Hampshire was the first state (in 1851) to regulate insurance, and every state now has an insurance commission, board, or department. Although the original purpose of insurance commissions was to protect consumers from fraud, more recently they have moved into other areas, such as requiring minimum benefits and developing regulations related to policy renewal and cancellation. Any policies concerning the use of genetic tests by insurance companies most likely would also be promulgated by regulation within the individual states.

Currently, an estimated 85 percent of the American population has health insurance coverage: 10 percent of the population has some sort of public coverage, and 75 percent is covered privately. In 1984, 80 percent of those with private coverage received their insurance through their employer, while approximately 15 percent purchased it individually. Among those with individual coverage, 64 percent were insured by a commercial company, 29 percent were insured by a nonprofit (Blue Cross/Blue Shield) plan, and 7 percent were covered by an HMO. Approximately 74 million people are insured by nonprofit plans, 93 million are insured by commercial health insurance companies, and 71 million are covered by HMOs or self-insured plans.

A major change in the health insurance industry in recent years is the trend by employers to self-insure. An employer who self-insures does not pay premiums to private insurance companies but instead accepts the risk itself. Employees file claims directly with their employer and are reimbursed out of the company's reserves. In many instances, insurance companies sell employers "stop-loss" coverage—essentially a catastrophic policy that protects employers from unexpectedly high individual or annual claims. Many employers also contract with organizations that administer and process their claims.

The Employee Retirement Insurance Security Act (ERISA) dictates that self-insurers are exempt from state insurance regulations. As such, they need not offer the minimum benefits mandated by a given state. In Wisconsin, for example, which in April 1992 became the first state to regulate the use of genetic tests by insurance companies, employers who self-insure are exempt from such regulation.

Employers have an incentive to self-insure both because of the savings achieved by avoiding insurance regulations and because they are able to select which health services will be covered. Results of a survey published in 1985 report that 58 percent of all employers self-insure. Given that these are mostly the large employers, an even greater percentage of employees are actually covered by self-insured plans.

Implications for Research

Genetically based tests could be used by insurance companies in one of two ways: companies could engage in direct testing—that is, they could require applicants or insureds to take a genetically based test as a criterion of eligibility; or they could engage in indirect testing, by requiring applicants or insureds to disclose the result of previous testing. Much of this testing is obtained through genetic research.

Whether insurers should be prevented from asking applicants about information acquired when they were research subjects is a difficult question ethically. From the insurer's point of view, it is vital that applicants enter an insurance agreement with no more information about their health than insurers have; private insurance relies on the absence of adverse selection. At the same time, allowing insurers to ask about information acquired through research erodes the public's trust of and, subsequently, participation in biomedical research. If insurers are permitted to discriminate against individuals who volunteer for research when others of equally high risk who did not volunteer are not penalized, public willingness to participate in research is certain to diminish. There are clear societal benefits to research, and if these are to be protected, there should no be tangible disincentives to participation. Particularly since the majority of biomedical research is supported through public funds, the argument can be made that research findings should be used for public rather than commercial benefit. Moreover, it is unethical to deny subjects information about their own health that was collected through the research process.

Given the difficulty of protecting the confidentiality of research subjects when research data is banked, researchers would be well advised to avoid banking whenever possible. However, avoiding banking imposes a scientific hardship, since it would be useful to try new tests on banked specimens, and there may be greater benefit in having certain identifying information accompanying the specimens. If it were clear that no third parties, including insurance companies, could gain access to researchers' data, scientific compromises of this sort would not need to be made.

Economic Implications

In any sort of risk screening by insurers two economic interests are directly at odds: those of the insurance industry and those of the public. Insurers "are businesses that are accountable to their policyholders and stockholders. They must generate a profit for those who have invested in the company." At the same time, most costs associated with care for individuals excluded from the private insurance market ultimately become the responsibility of the public, either directly, through government-sponsored programs, or indirectly, through compensation to hospitals for their "bad debt."

Insurers emphasize that risk classification allows them to set prices more precisely, which leads to lower-risk capital requirements and higher profits or returns for stockholders. It is argued that if insurers were limited in the degree to which they could classify risk, good risks would be forced to subsidize bad risks, and good risks then would choose to leave the group. Of course, such an occurrence could only happen if "good risks" had an alternative source of insurance that was still classified by risk.

Determining the moral rightness of this debate requires resolving whether insurance is a private business or a social institution. Insurers' arguments defending the need to classify risk (and, thereby, to screen for genetic conditions), are founded in the notion that risk classification makes good business sense. Insurers readily admit that such a system leaves certain people uninsurable, but do not believe that they should be held responsible for that problem. It is not the intent of the insurance industry to ensure access to health care for all, nor does the industry perceive its responsibility to be one of facilitating equality of outcome.

Typically, economics has guided insurance industry practices. Regulating private insurance has interfered with risk assessment procedures only when risk assessment involved legally prohibited categorical discrimination (for example, racial discrimination) or when a particular state decided that insurance procedures were unfairly discriminatory. Risk assessment rarely has been limited by regulation so as to make private insurance more accessible to the public.

Our nation has no consistent understanding of the degree to which private businesses have a responsibility to promote social justice. Employment discrimination based on physical handicap is legally prohibited, even where an employer must incur additional expenses—for example, by building a ramp—to comply with the law. In a similar vein, to broaden access to health care, certain regulations have been imposed on private health insurance companies, such as those requiring that specified minimum benefits be provided. However, such regulation remains piecemeal, and it does not address the broader question of whether companies have a responsibility to provide health care to the sickly or to those who cannot afford insurance.

Policy Options

It must be remembered that screening programs are not simply either mandatory or voluntary, but lie along a whole spectrum of possible programs. Among these is what might be called "conditionally mandatory" screening: to have access to a particular institution or benefit, one must consent to be screened, although one is not obligated to be screened generally. Requiring applicants for health insurance to be screened as a condition for eligibility could be thought of as an example of a conditionally mandatory policy. Given the ubiquity of insurance, requiring screening de facto means compulsory screening for most people. If instead public policy were guided by the President's Commission, which recommended that genetic screening be voluntary, insurers would not be allowed to screen.

The policy options that could be implemented to govern the use of genetic screening by insurance companies range from no limits of any kind to a complete prohibition against screening. A further option could be to enact legislation that would affect the provision of insurance without regulating companies directly. The policy options fall into four categories: federal insurance legislation; state insurance regulation; other federal legislation; and other statewide legislation.

Federal insurance regulation. Insurance companies are regulated on a state-by-state basis. Federal legislation, however, has been proposed that would fundamentally change how insurance is provided in this country. While it is not the thrust of this paper to conduct an analysis of the many recent proposals for national, comprehensive coverage, a brief description of their scope can be sketched, particularly as the various proposals would conduct risk screening or provide access to insurance for those with predispositions to genetically linked diseases.

Broadly speaking, two types of national health plans have been proposed as options for this country: plans that are completely publicly funded, and plans that have some balance of public and private involvement. Publicly funded plans either could establish a national health *insurance system,* as Canada has done, or a national health *service*, like the one in Great Britain. With either of these types of plans, risk screening does not exist. Membership in the plan is automatic, like public school enrollment in the United States, and no proof of eligibility is required. One is entitled to a place by virtue of being a citizen.

Historically, there has been too little political support in the United States for a completely publicly funded system to warrant the belief that one might be realized in the near future. Nonetheless, national health insurance (or a national health service) remains the only option that would spread risks across all individuals.

National health programs that mandate coverage for all persons, but endorse a mixed private/public system, have also been proposed. In most of these plans private coverage would remain an employee benefit, but public coverage would be available to and required of everyone whose employers did not provide coverage or who were ineligible for or unable to afford private coverage. In such a system, the risk screening practices of insurance probably would remain unchanged, but the government would ensure that no one was left uncovered.

State regulation of insurers. If insurance were to remain regulated on a state-by-state basis, several options for reform exist. The greatest protection for those with a genetic predisposition (or for anyone with any other medical risk or condition) would be to return to a community-rated system. Within such a system there would be no place for genetic testing, because applicants would not be rated according to their individual health risks or conditions. Insurance companies would fare no worse within a community-rated system than they do now, assuming all companies were required to set their premiums by community rating. The system would not allow healthy or low-risk individuals to band together to form groups that would pay less for health insurance, as permitting this would create a multi-tiered system that raises the rates significantly for those with higher risk conditions, in addition to forming an underclass of medically ineligible individuals.

A reform that would more directly respond to some of the concerns about insurers using genetic tests would be to leave the structure and practices of the current system intact, but to have state insurance commissions pass regulations prohibiting insurance companies from using genetic screening technology in particular for eligibility or rate-setting purposes. As risk screening is the foundation on which private insurance currently is based, prohibiting *all* risk screening will not be acceptable to companies. According to one insurance spokesperson, if risk classification were abolished, "equity" (that is, fairness) would give way to "equality" (that is, sameness), "and private insurance as it is known today might well cease to exist." However, insurers could be limited in the *degree* to which they classify risk. The pressure to limit risk screening in some way will only escalate as conditions that make an increasing number of persons a "substandard" or "deniable" risk become detectable.

Genetic testing may be an area around which insurers would be more willing to make compromises than others. For example, insurance spokespersons have suggested that while insurers must be able to perform HIV antibody tests, perhaps they could avoid conducting genetic screening. After all, actuarial classifications already incorporate the cost of treating genetic conditions since genetic disease always has been with us, while HIV disease is a new condition. Although new genetic testing technology could further refine risk assessment that would allow companies to charge higher premiums to individuals expected to make more claims, the screening would not identify cases that had not already been included in overall cost projections.

That HIV has affected socially vulnerable segments of the population to a disproportionate degree, whereas *everyone* is viewed as subject to genetic conditions, is a fact that also cannot be ignored. Genetic risks seem to occur more randomly than the risk for HIV, and this may influence industry or public policy responses. This is not to suggest that insurers will not want to use genetic tests, particularly as testing technology both improves and becomes less expensive. However, it is to say that insurers do not treat all types of risk assessment identically, and, whereas the industry would consider a blanket prohibition of assessment preposterous, certain specific limitations may be negotiable.

Another option would be to allow companies to screen, but to prohibit them from asking about information applicants might have learned on their own. This option would be in keeping with the position that insurers should be permitted to assess risk, but would not allow an individual applicant to be harmed for having followed a physician's recommendation to be tested or for having been a participant in research. Particularly if there are public health or public policy recommendations that all persons or certain subgroups be tested for genetic conditions, it is morally unacceptable essentially to punish those who have followed these recommendations by increasing their likelihood of being denied insurance when others of identical risk who have not chosen to follow the recommendations remain insurable.

Other federal legislation. Given that some might reject the seemingly anticompetitive nature of a completely community-rated system, a compromise between widespread community rating and our present system would be to make changes in the Internal Revenue Service Code to allow nonprofit plans once again to be tax-exempt, but to stipulate more carefully the conditions for qualifying. Such conditions would likely include open enrollment, community-rated premiums, and no exclusions for preexisting disorders.

Another important change that could occur on the federal level would be no longer to allow employers who self-insure to remain exempt from the regulations imposed on insurance companies. Private insurance companies advocate lifting the ERISA exemption for self-insurers, since the exemption places insurance companies at a competitive disadvantage. Companies that self-insure not only have greater freedom of movement, but can also offer fewer benefits (and, therefore, lower premiums) than insurance companies.

Other statewide legislation. Seven percent of the uninsured in this country are estimated to be "medically uninsurable"—that is, they have been unable to obtain health insurance due to their disease histories or preexisting conditions. Many people believe that policies should be created to provide greater access to health care for those at medically higher risk, but they also believe that it is not the responsibility of the private insurance industry to guarantee that access. In response to this sentiment, states may decide to enact programs designed to expand coverage to the uninsured within their own jurisdictions. Persons excluded from the private insurance market due to genetic risk could qualify for such programs.

One option implemented by some states has been the establishment of high-risk pools for those considered medically uninsurable. To qualify, an individual typically must have been rejected from an insurance company at least twice because of medical risk. Those enrolled in pools pay premiums, but as this group is the most expensive to insure, there is a cap on how high a premium the state can charge. Costs not covered by premiums are financed in part by state revenues and, in greater part, by insurance companies, which contribute to pools in proportion to their share of the market in a given state. But as risk pools comprise only the highest-risk individuals, it should come as no surprise that they have never been viable financially.

Another option would be to mandate comprehensive coverage. The states of Massachusetts and Oregon have passed legislation requiring all employers to provide health insurance to employees or to contribute to a fund out of which insurance for uninsured employees could be financed; similar initiatives are underway in other states.

Challenging Precedent

There is little doubt that a broader sharing of risks and less restrictive eligibility requirements would result in greater access to health insurance for individuals who have had difficulty securing a policy. It also is true that a broader sharing of risks would result in higher premiums for low-risk individuals. And yet in many of our other social arrangements we do not ask those in greatest need to shoulder their burdens alone. Services such as the national defense, fire protection, and education are financed by all, regardless of who benefits. We do not ask parents of a retarded child to pay higher education taxes or individuals who have been the victims of crime to pay more for police protection, and yet we do consider it appropriate to have our main health care payment system be one in which the sick (or those more likely to become sick) pay considerably more than those who are well, if they are not excluded from the system altogether.

If the laws and regulations governing the practice of insurance in this country do not change, the genetic testing that will be made possible as we continue to map the human genome may result in many more individuals being denied private insurance coverage than ever before. According to our present system, there is nothing wrong with this. Indeed, insurance companies accurately defend themselves by claiming that they would be treating genetic conditions exactly as they treat other conditions. What is new as a result of genetic research is the vast number of people who would be affected by insurance company exclusions. Precedent exists for insurance companies to classify applicants by risk and to make exclusions accordingly. However, precedent also exists for insurers and, certainly, for other private businesses to be regulated when there are overriding social or public policy concerns. Insurance occupies an integral place in providing for the welfare of the majority of the population. As such, the Human Genome Project will challenge us to weigh the interests of private businesses against the interests of our citizenry in a responsible manner. Following precedent is not the only policy option available.

PRIVACY RULES FOR DNA DATABANKS[4]

The new genetics promises more precise diagnostic capabilities, a greater understanding of how genetic factors influence disease, and eventual treatment and prevention strategies. But unless we agree on and enforce reasonable privacy rules related to genetic information, the price of using the new genetics in medicine may be privacy itself. The genetic information gleaned from an individual's DNA is similar to information contained in medical records. However, the DNA molecule itself holds much currently undecipherable information and may usefully be analogized to one's "future diary" written in a code we have not yet broken. Storing DNA molecules, or probablistic future diaries, in a "DNA bank" for future analysis presents novel privacy issues that merit widespread discussion and early action. When the DNA bank contains information derived from the DNA sample, it becomes a "DNA databank." James Watson has said, "The idea that there will be a huge databank of genetic information on millions of people is repulsive." Why is such a databank repulsive, and what action can effectively safeguard the genetic privacy of individuals in our coming genetic age?

Increasingly precise genetic information has the potential to alter radically our views of privacy because control of and access to the information contained in an individual's genome gives others potential power over the personal life of the individual by providing a basis not only for counseling, but also for stigmatization and discrimination. Genetic information also has unique pri-

[4]Reprint of article by George J. Annas from the *Journal of the American Medical Association* v 270 pp2346–50 N 17 '93. Copyright © 1993 by The American Medical Association. Reprinted with permission.

vacy implications, since genetic information is immutable and the DNA molecule is stable and will also provide information about the individual's parents, siblings, and children. Finally, genetic information has been grossly misused in the past, especially in the eugenics movement and Nazi Germany's program of racial hygiene. The uniqueness of genetic information, coupled with computer technology and a general distrust of large, bureaucratic record-keeping systems, requires credible privacy controls if DNA databanks are to be permitted. Current policies and practices governing the privacy and confidentiality of medical information are woefully inadequate to protect personal privacy and liberty in the new genetics age. Therefore, new rules for DNA databanks are needed now to help minimize the harm to individual privacy and liberty that the collection, storage, and distribution of genomic information could produce, and to permit socially useful medical and epidemiologic applications of genetic information. As the US House of Representatives Committee on Government Operations concluded in its study of genetic information, such rules "will be more effective and less expensive to implement if established in advance." This article summarizes the major privacy issues at stake in DNA databanks and suggests a set of tentative rules to begin a serious dialogue on how best to protect genetic privacy.

Law and Practice Involving Medical Records

Since genetic information is most analogous to medical information, it is useful to begin with a summary of current protections of medical records. More than 15 years ago Justice William Brennan, concurring in a case upholding the constitutionality of a New York law that required the storage of drug prescriptions in a central computer (for the purpose of identifying prescription misuse), expressed his growing concern over the privacy implications of computerized medical information: "The central storage and easy accessibility of computerized data vastly increase the potential for abuse of that information, and I am not prepared to say that future developments will not demonstrate the necessity of some curb on such technology." That time has arrived. There have been major changes in computerization that make medical records more accessible without corresponding changes in legal protections.

Almost all of the law dealing with access to medical records by persons other than the patient can be categorized under the headings of confidentiality, privilege, and privacy. *Confidentiality* presupposes that something secret will be told by someone to a second party (such as a physician) who will not repeat it to a third party (such as an employer). In the physician-patient context, confidentiality is understood as an expressed or implied agreement that the physician will not disclose the information received from the patient to anyone not directly involved in the patient's care and treatment.

A communication is *privileged* if the person to whom the information is given is forbidden by law from disclosing it in a court proceeding without the consent of the person who provided it. Privilege, sometimes called testimonial privilege, is a legal rule of evidence, applying only in the judicial context. The privilege belongs to the patient, not to the professional, although the hospital, physician, or databank may have a duty to assert it on behalf of the patient.

There are at least four senses in which the term *privacy* is generally used. The first three describe aspects of the constitutional right of privacy. The central one, found in the liberty interests protected by the 14th Amendment, is the right of privacy that forms the basis for the opinions by the US Supreme Court limiting state interference with intimate, individual decisions, such as those involving birth control and abortion. The second and third types of constitutional privacy protect certain relationships, such as the husband-wife, parent-child, and physician-patient relationships, and certain places, such as the bedroom, from governmental intrusion.

There is also a fourth sense of privacy, the common-law right of privacy, that applies to private actions: "the right to be let alone," to be free of prying, peeping, and snooping, the right to keep personal information inaccessible by others.

It has become common to use the phrase "data protection" to describe informational privacy, especially in the realm of medical records, financial records, employment records, and criminal records. Nonetheless, the phrase is inadequate in the case of genetic information because three of the four types of privacy (informational privacy, relational [family] privacy, and decision-making privacy) overlap. This, combined with DNA's coded "future diary" character, create unique concerns regarding invasion of privacy.

Concerns about informational privacy have historically centered on the use to which authoritarian governments can put personal data to control the lives of individual citizens. Thus,

Orwell's vision of "Big Brother" in *1984* provided a powerful call to action, as did Aleksandr Solzhenitsyn's vision of authoritarianism in *Cancer Ward:*

> As every man goes through life he fills in a number of forms for the record. . . . A man's answer to one question on one form becomes a little thread, permanently connecting him to the local center of personnel records administration. There are thus hundreds of little threads radiating from every man. . . . They are not visible, they are not material, but every man is constantly aware of their existence. . . . Each man, permanently aware of his own invisible threads, naturally develops a respect for the people who manipulate the threads . . . and for these people's authority.

The loss of individuality and control over one's life has been a theme of data protection, as have the horror stories of false data being stored in one's government profile. However, DNA databanks will be developed not only by governmental agencies, but also by private corporations, hospitals, and physician researchers. It is thus useful to examine both governmental and private DNA databanks. Moreover, we will be equally concerned both with protecting accurate information and preventing false information. It will also be important to distinguish between collection and storage of the DNA sample itself, and storage and distribution of information derived from analysis of the DNA sample, although most DNA databanks will likely do both. Currently, DNA "fingerprinting" for criminal suspect identification provides the most important example of a governmental DNA databank.

DNA Fingerprinting

A suspect may be placed at the scene of the crime in a number of ways. The most common is by an eyewitness. But eyewitnesses are notoriously unreliable, and most prosecutors prefer to have eyewitness testimony supplemented by physical evidence, such as fingerprints or footprints. In violent crimes such as rape and murder, perpetrators may leave sperm or blood behind, or carry away some of the victim's blood on their person. By means of ABO blood groups, individuals can be excluded as suspects, because their blood types do not match the sample left at the scene of the crime. With the use of DNA profiles, however, suspects can not only be excluded but can also be identified as the source of blood or sperm.

The treatment of DNA fingerprinting by the US courts provides insights as to how courts may deal with genetic information in general. Courts in the United States already agree that DNA fingerprinting is scientifically accepted, although there have been disputes over the proper evidentiary standard for admitting DNA fingerprinting for use by the jury in a criminal trial. Both a 1992 Circuit Court of Appeals case and a 1992 report of the National Research Council recommended that DNA evidence be admitted into evidence and that continuing statistical disputes about determining the probability of a match be explained to the jury. The June 1993 US Supreme Court opinion regarding the admissibility of scientific evidence, although it did not deal directly with DNA fingerprinting, permits scientific evidence to be given to the jury as long as the judge finds it is both relevant and "reliable." Law and science seem to be on the same track here, and the stage is set for an explosion of DNA databanks for criminal law purposes. [Editor's note: The issue of DNA fingerprinting is discussed more fully in the sixth article in this section.]

Some states now condition parole on the deposit of a DNA sample with the police (in most cases for sex offenders only, but it seems likely that all felonies will soon be included, and possibly all crimes). The purpose, of course, is to use the sample to construct a DNA "fingerprint" to "identify" the perpetrator of current sex crimes among former sex offenders. Two early court decisions suggest that convicted felons have no reasonable expectation of privacy that would prohibit states from requiring them to give a blood sample to be used for producing such a DNA fingerprint to be stored in a criminal record-keeping system. One court could see no difference in privacy concerns between a DNA fingerprint used for identification and a traditional fingerprint. It could be seen as reasonable at some point in the future to have the Federal Bureau of Investigation store DNA fingerprints (just as the Bureau now has a large proportion of the population's fingerprints) to make the job of law enforcement easier. The central problem is that this treats everyone in the United States (whose DNA is on file) as a crime suspect, making us a "nation of suspects," and radically alters the relationship between the citizen and government.

The more generic DNA databank problem is that once a governmental agency has a DNA sample, it can learn much more about the individual than just whether or not their DNA profile "matches" that taken from a crime scene. The agency not only can discover the genetic makeup of an individual, but also, in the future, may be able to learn about genetic predispositions—the probability of an individual developing specific genetically determined or genetically influenced diseases. There are currently no standards for such criminal DNA databanks. This is one reason the Canadian Privacy Commissioner has recommended that "Governments should not establish banks of genetic samples [or genetic databases] of convicted persons or the general population for criminal justice purposes.

DNA as a Future Diary

Both governmental and private DNA databanks contain information that may be considered unique and significantly more personal and private than either fingerprints or medical records. A medical record can be analogized in privacy terms to a diary, but a DNA molecule (as distinguished from information already derived and recorded from the DNA sample) is much more sensitive. It is in a real sense a "future diary" (although a probabilistic one), and it is written in a code that we have not yet cracked. But the code is being broken piece by piece such that holders of a sample of an individual's DNA will be able to learn more and more about that individual and his or her family in the future as the code is broken. Of course, such predictions will not be precise, because the expression of genetic characteristics will vary from never expressed to expressed in an extreme manner, and individuals with the same somatic expression of a genetic condition will respond differently. Nonetheless, this is information individuals, physicians, insurance companies, employers, and others will want and on which they will base decisions affecting the individual. Health insurance companies, for example, may wish to deny coverage to applicants with genetic susceptibilities to serious, potentially expensive, diseases.

Medical researchers and epidemiologists will want access to the information stored in large DNA databanks to search for genetic connections to disease. One can even envision law enforcement or child protection agencies looking for children with genetic conditions to make sure their parents are providing them with proper medical care or prevention strategies. Although it seems far-fetched today, assuming that a gene that predisposes a person to skin cancer is discovered in the future and that such cancer is preventable if one stays out of the sun, it would be possible to search DNA databases to identify these individuals and counsel them about their risks. If this is seen as reasonable, the next step would be to identify children at risk and require that their parents protect them from this genetic hazard by keeping them out of the sun and away from the beach. Similar scenarios can be constructed for virtually all diseases that have genetic predispositions. Although similar arguments can be made for high blood pressure and high cholesterol levels, the identification of a genetic link will make prediction seem more scientific and thus prevention seem more urgent.

It seems reasonable to conclude that the mere existence of the technology to "decode" DNA will lead us to alter radically our view of informational privacy. In the past we have put special emphasis on information that is potentially embarrassing and sensitive (such as sexually transmitted diseases) and on information that is uniquely personal (such as a photograph of one's face). Genetic information is both potentially embarrassing and uniquely personal. The existence of such decodable information could either impel us to take privacy much more seriously in the genetic realm than we have in the medical and criminal realms, or lead us to give up on maintaining personal privacy altogether. This latter response seems defeatist and unlikely, although one leading medical geneticist has already suggested that "we must prepare for others to know." Is this true, or can genetic privacy be protected? The issue of genetic privacy revolves around choice in discovering and exposing personal genetic information. To return to the closest analogy, what lessons can we learn from the rules governing current medical information systems that might help us maintain control over our genetic information?

Rules for Medical Information Systems

Rules about medical information are mostly state rules, and there has been little serious study of the privacy of medical record-keeping systems since the early 1970s. At that time, when computerization of medical records was in its infancy, the US Congress passed the Privacy Act of 1974, which, among other things, established the Privacy Protection Study Commission (instead of a permanent regulatory commission). The commission's 1977 report remains the most thoughtful and authoritative statement on large record-keeping systems. In regard to medical records, the commission found that medical records contain more information and are available to more users than ever before; that the control of health care providers over these records has been greatly diluted; that restoration of this control is not possible; that voluntary patient consent to disclosure is generally illusory; that patients' access to their records is rare; and that there are steps that can be taken to improve the quality of records, to enhance patients' awareness of their content, and to control their disclosure. Some of the commission's major recommendations were as follows.

1. Each state should enact a statute creating individual rights of access to, and correction of, medical records, and an enforceable expectation of confidentiality for medical records.

2. Federal and state penal codes should be amended to make it a criminal offense for any individual knowingly to request or obtain medical record information from a medical care provider under false pretenses or through deception.

3. On request, an individual who is the subject of a medical record maintained by a medical care provider, or another responsible person designated by the individual, should be allowed to have access to that medical record, including the opportunity to see and copy it; and have the opportunity to correct or amend the record.

4. Each medical care provider should be required to take affirmative measures to ensure that the medical records it maintains are made available only to authorized recipients and on a "need-to-know" basis.

5. Any disclosure of medical record information by a medical care provider should be limited only to information necessary to accomplish the purpose for which the disclosure is made.

6. Each medical care provider should be required to notify an individual on whom it maintains a medical record of the disclosures that may be made of information in the record without the individual's express authorization.

Of course, it is not just the storage of information that is problematic. It is the use of such information by third parties to make decisions about the future of individuals that puts an individual's privacy and liberty interests most directly at risk. The Privacy Commission was careful to specify rules for the release of data identified with a particular individual from the databank. The Privacy Commission discovered, for example, that often when an individual applies for a job, life or health insurance, credit or financial assistance, or services from the government, the individual is asked to relinquish certain medical information. Although this is necessary in many cases, the commission found that individuals are generally asked to sign open-ended or blanket authorizations with clauses such as one requiring the recipient to "furnish any and all information on request." The commission accordingly made the following recommendations.

Whenever an individual's authorization is required before a medical care provider may disclose information it collects or maintains about the individual, the medical care provider should not accept as valid any authorization that is not (1) in writing; (2) signed by the individual on a date specified or by someone authorized in fact to act on the individual's behalf; (3) clear as to the fact that the medical care provider is among those either specifically named or generally designated by the individual as being authorized to disclose personal information; (4) specific as to the nature of the information the individual is authorizing to be disclosed; (5) specific as to the institutions or other persons to whom the individual is authorizing information to be disclosed; (6) specific as to the purpose(s) for which the information may be used by any of the parties named in criterion 5 both at the time of the disclosure and at any time in the future; and (7) specific as to its expiration date, which should be for a reasonable time not to exceed 1 year.

It is somewhat remarkable that these recommendations have not been generally adopted, and that the attention focused on privacy issues in the 1970s virtually disappeared in the 1980s. We may, however, be witnessing a renewed interest in the privacy of medical records. President Clinton's Health-Care Reform plan, for example, acknowledges that "currently, no uniform, comprehensive privacy standards related to health care information exist." His draft plan includes a separate section on "protection of privacy" that calls for, among other things, the establishment of "national privacy safeguards covering all health records, based on a Code of Fair Information Practices," and the creation of a "Data Protection and Security Panel" under the proposed National Health Board to set national "privacy and security standards through interpretative rules and guidelines" and to monitor and evaluate "the implementation of standards set by statute, regulations and guidelines." The recommendations of the Privacy Commission are a reasonable place to start the standard-setting process.

These recommendations and similar data protection principles suggest that there should be even more stringent rules for the collection, storage, and distribution of DNA molecules and information derived from them because of the unique characteristics of genetic information, including the fact that DNA molecules contain an individual's probabilistic future diary. Entities that can properly be described as DNA databanks already exist, and others are currently in the process of development. These include state phenylketonuria blood sample storage facilities; the Federal Bureau of Investigation's program to store DNA samples from convicted sex offenders and others; the US Army's DNA sample storage program; private genetic research projects, many involving pedigree research; the Red Cross and other blood donor and donor deferral programs; and private sperm, ovum, and embryo banks. Databanks developed for different purposes may require additional privacy safeguards, but all DNA databanks raise fundamentally similar privacy concerns. Some commercial DNA databanks already have policies in place, but the current state of the art of DNA sample and data protection is both spotty and rudimentary.

Privacy Rules for DNA Databanks

Since there are no existing privacy rules for DNA databanks, and since most genetic samples are now being collected and stored by either hospital-based programs or private clinics and corporations, it seems prudent to suggest a moratorium on such storage until reasonable rules are developed. On the other hand, because most storage (outside of military and law enforcement agencies) is currently in private hands, it seems unlikely that any agreement on a moratorium could be enforced without federal legislation. An alternative approach is to develop uniform state legislation, although the prospects for a uniform law seem remote. Some states, including Wisconsin, Florida, and Arizona, have already enacted legislation designed to curb discrimination on the basis of genetic information. These laws are similar to state laws previously enacted to prevent discrimination on the basis of human immunodeficiency virus infection. Florida's statute also prohibits DNA analysis (outside the criminal law context) without informed consent and further prohibits disclosure of the results of DNA analysis without consent. Violation of these prohibitions is a criminal offense. In addition, the person who performs the DNA analysis or receives the results must provide the person tested with a notice that the test was done and the results obtained, and on request by the person the results will be made available to the individual's physician (but not directly to the individual).

Since neither uniform state laws nor early effective federal legislation seem likely, it is probably more constructive to try to get voluntary agreement on the rules all DNA databanks should follow with or without legislative mandate. Legislation will, however, ultimately be needed to provide for criminal penalties for willful breaches of privacy. The following preliminary rules are suggested to protect individual privacy while permitting responsible medical research and treatment goals to be pursued.

1. No DNA databank should be created or begin to store DNA samples until there is (*a*) public notice that the DNA databank is to be established, including the reason for the bank, and (*b*) a privacy impact statement prepared and filed with a designated public agency that is also responsible for developing and enforcing privacy guidelines for the DNA bank (ultimately, a DNA databank-licensing board should be established to license all DNA databanks in the United States with uniform rules); (*c*) the burden of proof should be on the DNA bank to establish that storage of DNA molecules is necessary to achieve an important medical or societal goal.

2. No collection of DNA samples destined for storage is permissible without prior written authorization and agreement that (*a*) sets forth the purpose of the storage; (*b*) sets forth all uses, including any and all commercial uses, that will be permitted of the DNA sample; (*c*) guarantees the individual (i) continued access to the sample and all records about the sample, (ii) the right to correct inaccurate information, and (iii) the absolute right to order the identifiable sample destroyed at any time; and (*d*) guarantees the destruction of the sample or its return to the individual should the DNA databank significantly change its identity or cease operation.

3. DNA samples can be used only for the purposes for which they are collected, and linkages to other computerized information systems are prohibited. Specifically, there may be (*a*) no waivers or boilerplate statements that permit other uses; (*b*) no access to the DNA information by any third party without written notification to the individual whose sample is being used; (*c*) no access by third parties to any personally identifiable information; and (*d*) strict security measures, including criminal penalties for misuse or unauthorized use of DNA information.

4. Mechanisms should be developed to notify and counsel those whose DNA samples are in storage when new information

that can have a significant health impact on the individual is obtainable from their stored DNA sample.

Comment

Most of these proposed rules are self-explanatory. It may seem premature to develop rules or guidelines for DNA databanks, but the long history of medical record keeping, the short history of DNA fingerprinting, and the intermediate history of sperm banking have demonstrated that standards are necessary. Some questions these proposals have raised merit comment. First, where is the "designated public agency" responsible for developing and enforcing privacy guidelines? The response, that there currently is no such agency, is not satisfying. There should be one and under any reasonable national health care program there will be one. It should be a federal agency because few, if any, DNA databanks will operate solely within the confines of any one state. This agency should be as independent as possible from the funding agencies (such as the National Institutes of Health) and should probably be located either in an existing agency, like the National Institute for Standards and Technology, or created as an independent agency.

The requirements for "informed storage" in part 2 are not remarkable and are analogous to both medical record storage and embryo storage. Embryos are, of course, even more important than DNA samples, since they have the potential to become children. Typical storage contracts currently require the couple to agree to such things as disposition of the embryos on the separation, divorce, or death of one or both of the couple, as well as limiting the terms of the storage and providing for other contingencies. Even more elaborate storage agreements are used when an individual wants his or her entire body frozen and stored for possible "treatment" at some distant time in the future. The point is not that we should treat DNA samples like cryopreserved embryos or bodies, but that detailed storage contracts and consent forms are not a novel idea and can be implemented.

Part 3 is relatively standard privacy protection language, although many researchers and commercial enterprises might object to keeping track of current addresses and to a requirement (3*c*) forbidding all third-party access to identifiable information. In the research context, the practice has been to seek institutional review board approval for such uses, but this is very unsatisfactory. The institutional review board did not approve the new use when the sample was collected and, of course, the individual could not have given his or her consent unless the type of research was disclosed and agreed to at the time the sample was collected. Obviously, this agreement cannot be generic (eg, "all genetic research"). It could, however, include specific goals of research even though all specific means are not currently envisioned (eg, "all attempts to locate cystic fibrosis genes").

Part 4 is the vaguest rule and the one that requires the most work to make operational. There is legal precedent for holding physicians responsible to recontact patients when the physician learns of a new danger related to previous treatment. On the other hand, DNA banks will not always be run either by physicians or for therapeutic ends. In addition, since most new genetic tests first appear in the nation's television broadcasts and daily newspapers, notification may be less important than counseling options. Each DNA databank could also have a newsletter that it routinely sends to all "depositors," and new direct genetic tests or treatments could be described therein. It seems unlikely that the "duty to protect" the depositor's family members will ever arise at a DNA bank, although research facilities may discover a genetic condition, such as susceptibility to breast cancer, that has serious implications for family members. If this is a reasonable possibility, the facility's policy on disclosure to genetically related family members should be detailed at the time of deposit, so the individual who disagrees with it can keep his or her DNA sample out of the bank. Whether and to what extent actual genetic testing should be financed by any system of national health care is a policy question that merits a separate analysis.

This is a preliminary proposal that requires additional discussion, debate, and refinement. Criminal and military DNA banks may even be an exception to many of these rules, but only if the samples are converted into DNA profiles that are used for identification purposes only. Compromises in the security of the data contained in the DNA molecule will ultimately compromise the privacy and liberty of the individuals whose DNA is stored in DNA databanks. Scientists, physicians, and the public should take a strong stand in favor of privacy and against the establishment of a genetically based surveillance society.

USES AND ABUSES OF HUMAN GENE TRANSFER[5]

. . . Human genetic engineering has been divided into four categories: (i) somatic cell gene therapy (correcting a genetic defect in the somatic, or body, cells of a patient), (ii) germ-line gene therapy (correcting a genetic defect in the germ, or reproductive, cells of a patient so that offspring of the patient would be corrected), (iii) enhancement genetic engineering (insertion of a gene to try to "enhance" or "improve" a specific characteristic, for example, adding an additional copy of a growth hormone gene to increase height), and (iv) eugenic genetic engineering (insertion of genes to try to alter or "improve" complex human traits that depend on a large number of genes as well as extensive interactions with the environment, for example, intelligence, personality, character, *etc*).

Somatic cell gene therapy is now successfully underway in human patients and appears to have widespread public acceptance. Germ-line and enhancement genetic engineering are being carried out in laboratory and farm animals. Consequently, the technical ability to attempt human application of these latter techniques exists. It is critical, therefore, that serious ethical reflection by both experts and the general public begin on germ-line and enhancement genetic engineering. . . .

. . . When I initially coined the terms for the human genetic engineering categories (for testimony before then-Congressman Gore's Hearings on Human Genetic Engineering in 1982), I intentionally used emotive words to psychologically "draw a line." Thus, I used somatic cell and germ-line gene *therapy,* but *enhancement* and *eugenic* genetic *engineering*. "Therapy" is good, but "enhancement," "eugenics," and "engineering" stir up concern. . . .

As LeRoy Walters pointed out (Walters, 1985), there are really five, not four, categories because enhancement genetic engineering must be subdivided into somatic cell or germ line. And here is

[5]Reprint of editorial by W. French Anderson from *Human Gene Therapy* v 3 pp1–2 '92. Copyright © 1992 by Mary Ann Liebert, Inc. Publishers. Reprinted with permission.

the core of the concern about germ-line gene therapy. Few would argue against therapeutic germ-line therapy, but it is the slippery slope leading to attempts at germ-line enhancement that causes all of us to question whether a strict prohibition at the germ-line might not be the safest course in order to prevent serious abuses of human gene transfer. As powerful as this argument is, however, I still believe that we as caring human beings have a moral mandate to cure disease and prevent suffering whenever possible (and not just in humans but in animals also). Therefore, I strongly support germ-line gene therapy, but I have defined criteria that I believe are required to be met before attempting this type of therapy (see below).

There are those who, after long and serious contemplation, disagree with this position. They fear that the slippery slope has no stopping points, that enhancement engineering leading right on to brazen eugenics will occur if we let ourselves begin. I respect this viewpoint, but I have a fundamental faith in the wisdom and power of a well-informed public. Consequently, I believe that it is critical that extensive public discussions begin and that all the risks as well as all the potential benefits of germ-line gene transfer be aired. I support (and I believe that over time the public will support) germ-line gene therapy, but only if therapy can be clearly distinguished from enhancement engineering.

My reasons for objecting to enhancement engineering even at the somatic cell level are detailed in Anderson (1989, 1990). Somatic cell gene therapy for the treatment of severe disease is considered ethical because it can be supported by the fundamental moral principle of beneficence: it relieves human suffering. It is, therefore, a moral good. Under what circumstances would human genetic engineering not be a moral good? In the broadest sense, when it detracts from, rather than contributes to, human dignity. The justification for drawing a line is founded on the argument that, beyond the line, human values that our society considers important for human dignity would be significantly threatened.

I believe that somatic cell enhancement engineering would threaten important human values in two ways: first, it could be medically hazardous (*viz.*, the risk could exceed the potential benefits and could, therefore, cause harm), and second, it would be morally precarious (*viz.*, it would require moral decisions that our society is not now prepared to make and which could lead to an increase in inequality and an increase in discriminatory practices). . . .

If germ-line gene therapy for the prevention of serious disease is accepted as a moral good, perhaps even as a moral mandate, what criteria need to be met before initiating this potentially perilous procedure? I believe that there are three (discussed in detail in Anderson, 1985): (i) There should be considerable experience (over a number of years) with somatic cell gene therapy that clearly establishes the effectiveness and safety of treatment of somatic cells; (ii) There should be adequate animal studies that establish the reproducibility, reliability, and safety of germ-line therapy, using the same vectors and procedures that would be used in humans; and (iii) There should be public awareness and approval of the procedure since unborn generations will be affected. The gene pool is a joint possession of all members of society. Since germ-line therapy will affect the gene pool, the public should have a thorough understanding of the implications of this form of treatment. Only when an informed public has indicated its support, by the various avenues open for society to express its views, should clinical trials begin.

Gene therapy, somatic cell and germ line, has the potential for providing mankind with enormous good. It is imperative that we do not deny ourselves, our children, and our children's children the benefits from this technology simply because we fear potential misuse. It is our responsibility to establish adequate safeguards to prevent the abuses so that the full power of gene therapy can be realized and our children can live longer and healthier lives.

DNA GOES TO COURT[6]

. . . In the 1980s DNA and the technologies to manipulate and analyze it became new frontiers for patent and copyright law. DNA methods were also used to link suspects to crime, mainly rape and murder, thus drawing the pristine science of genetics into the courtroom battles between prosecutors and defense at-

[6]Excerpt of chapter 19 from *THE GENE WARS: Science, Politics, and the Human Genome* by Robert M. Cook-Deegan, pp299–306 '94.

torneys. Adapting genetics to social functions through law required accommodation on both sides.

In November 1983, residents of Leicester County, England, found the dead body of fifteen-year-old Lynda Mann by a path. She had been raped and killed by an unknown assailant. Traditional forensic methods were used, but the case was still not solved when fifteen-year-old Dawn Ashworth, from a nearby town, was also found raped and murdered in late July 1986. Richard Buckland, a worker at a local psychiatric facility, was arrested. The police attempted to link Buckland to the victims, and contacted geneticist Alec Jeffreys of Leicester University.

Jeffreys was a world figure in the development and analysis of human genetic markers. He was interested by the request and agreed to help out. He used DNA typing on material from vaginal swabs of the victims and compared them to suspect Buckland's DNA. Jeffreys concluded that the two young women had indeed been raped by the same man, but it was not Buckland. Buckland was released, despite having made a dramatic confession. Buckland became the first person exonerated by DNA testing.

The police then began a "genetic sweep" of the population in January 1987, intending to determine the DNA type of all young males in the vicinity. Colin Pitchfork was scared. He resorted to subterfuge, enticing a coworker to substitute for him when blood samples were drawn, so Pitchfork's DNA was not typed. By May, more than 3,600 DNA typings had been performed, but there was still no match to samples taken from the victims. In August 1987, a coworker admitted having substituted for Pitchfork, and six weeks later the police were notified. Pitchfork confessed to both murders and was convicted.

DNA typing had earlier been used to establish relatedness among individuals (to resolve disputed paternity, to enforce child support, or to allow immigration into the United Kingdom), but the Leicester County case was more widely publicized and promised far broader application to forensic testing. The Leicester case inaugurated a new technology for identifying individuals in criminal proceedings and touched off a battle that raged for several years. By the end of 1989, DNA typing had been used in at least eighty-five cases in thirty-eight states in the United States alone, and in spring 1991, every state had used forensic DNA testing.

For prosecutors, DNA typing was especially effective in rape and murder cases. DNA typing might enable them to link criminals to the scene of the crime as reliably as standard fingerprinting, but without requiring that the criminal leave a good fingerprint. In most crimes, there was a struggle, with bloodstains to be analyzed, semen from a rapist, or hair or skin tissue inadvertently left behind. If the perpetrator did leave behind a bit of hair, blood, semen, saliva, urine, or other tissues that could be typed by DNA analysis, then they could be identified. For defense lawyers, DNA typing could be an overwhelming exculpatory technology if there was no match. As in the Leicester case, DNA typing was more convincing than a false confession.

The power of the technology came from linking a person to the scene, not proving that the defendant committed a crime. (In rape, for example, a match between suspect and semen type indicated that intercourse took place, but not that it took place against the victim's will.) DNA typing was clearly a powerful new tool for law enforcement, but important questions remained about how to use it properly.

The techniques were quite similar to those used in pedigree studies for genetic linkage, and indeed used many of the same reagents. There were major differences, however, that emerged as more cases appeared in court. In genetic linkage, a genetic marker is followed through a family. The value of a marker is that it enables one to trace inheritance from one generation to another, to discern which marker was inherited from the mother and which from the father for each member of a pedigree. Typically, in pedigree research, fresh blood samples are taken from those in the pedigree.

In forensic investigations, however, there is less information to start with, since there are no family ties to help interpret the DNA findings. The critical question to answer is: How likely is it that these tissues—such as blood, hair, or semen—came from this particular suspect? The amount of material may be quite small, the sample is unlikely to be fresh, and it may be mixed with tissue from other individuals, as in cases of multiple rape, or mixed samples of both perpetrator and victim. Blood samples are often dried, and the DNA may be partially degraded. Sample DNA may be completely used up in the analysis, precluding reanalysis if the test fails and eliminating the possibility of further tests if initial tests are not definitive. Tests must be performed adequately by the laboratory, so that samples are not switched and criteria for calling a match are reliable.

Most important, however, one is not merely comparing DNA

type between parent and child within a pedigree, but rather trying to assess the likelihood that it comes from a particular person, the suspect. That assessment, in turn, depends on how often that DNA typing pattern occurs in the entire population, not just in a family pedigree. The probability of a match thus depends on statistical analysis of DNA typing patterns across the population and knowledge of how prevalent a given DNA type is. The statistical power of DNA typing thus ultimately rests on data that are expensive to collect, requiring systematic survey of the population with DNA typing of very large numbers of individuals.

As DNA typing entered the courtroom, questions about how adequately it had been performed and interpreted began to arise. The process of introducing the new evidence hinged on satisfying the *Frye* standard, a set of legal criteria that grew out of a 1923 murder case. A court was faced with deciding whether to admit evidence taken from one James Alfonso Frye, a young African-American accused of having murdered a white man in Washington, D.C. The prosecution proposed to introduce into evidence data about how his blood pressure responded to questions about the crime, as a measure of his veracity in a primitive precursor of the polygraph test. The court agonized, but ultimately rejected the proffered evidence, noting:

> Just when a scientific principle or discovery crosses the line between experimental and demonstrable stages is difficult to define. . . . while courts will go a long way in admitting expert testimony deduced from well-recognized scientific principle or discovery, the thing from which the deduction is made must be sufficiently established to have gained general acceptance in the particular field in which it belongs.

The court thus established a two-tiered sociological standard for the acceptance of scientific evidence. A court must decide the field whence it arose, i.e., identify a scientific community, and must determine that it was accepted within that community. These criteria were bulwarks against admitting evidence from new scientific techniques until the Supreme Court cast down the Frye standard in 1993. The reason for special caution was a belief that scientific data might unduly sway judges and juries. The *Frye* criteria, however, were rather vague. Just how to define a field and how to assess consensus was far from clear. The alternative to the *Frye* standard, under federal rules of evidence, was premised on relevance to the matter at hand—admitting into evidence anything that helped the court to assess the facts. The federal rules were developed under precepts outlined in a 1975 statute and subsequent amendments. Here also, while not so rigid as the *Frye* criteria, the judgment of admissibility turned on whether expert testimony would be useful in ascertaining or understanding the facts and required a judgment of the qualifications of experts. Rule 702 specified that expert status could be inferred from "knowledge, skill, experience, training, or education."

In the United States, most early DNA typing was performed by two private firms. One, GenMark, was affiliated with the British chemical giant ICI, and licensed the methods developed by Alec Jeffreys. Lifecodes was a small, independent firm based in Valhalla, New York. These private firms initiated forensic typing on a fee-for-service basis for prosecutors or defense attorneys.

DNA tests were first brought into American courts in 1986, but really caught hold in 1988. The Federal Bureau of Investigation began to focus on the promise of DNA testing as the technology was introduced into courtrooms throughout the states. The FBI set up a laboratory in Quantico, Virginia, to perform tests on request and to train those who wished to learn about the technology from state crime laboratories. The FBI also proposed to standardize the methods used so that results could be compared from one state to another and DNA typing profiles could be matched at the federal level, comparing samples analyzed by laboratories in different states. A California investigation might turn up a match to a Colorado serial killer, for example, using only the limited data from a computer code for DNA typing. The FBI could then notify police in both states to contact one another to pool their evidence.

As work on the OTA [Office of Technology Assessment] report on the genome project was winding down early in 1988, the number of criminal cases using DNA forensics rose quickly. It became clear that an assessment of forensic typing could also be useful. OTA thus began an assessment that produced a separate report in July 1990. The National Research Council of the National Academy of Sciences also formed a committee to assess forensic uses of DNA. The committee, chaired by Victor McKusick, began work in January 1990 and released its report in April 1992.

DNA evidence was first accepted as evidence in *Florida v. Andrews;* Tommy Lee Andrews was accused of having raped and slashed several women. DNA typing was used by the prosecution, and he was convicted in November 1987. In October 1988, the Florida State Court of Appeals for the Fifth District upheld the admission of DNA typing evidence. The prospects for DNA forensics looked rosy, but then some sloppy work showed how it could be troublesome.

The watershed case that cast doubt on how well DNA forensic testing was being performed was the highly publicized *New York v. Castro.* This case was tried in the same court caricatured in Tom Wolfe's novel *Bonfire of the Vanities.* Lifecodes had performed the DNA typing, concluding that a bloodstain on Jose Castro's watchband matched the blood types of a woman and daughter murdered in the building where he was a janitor. Lifecodes claimed that the likelihood of the match they found was one in 738 trillion. When expert witnesses scrutinized the evidence, it turned out that Lifecodes had ignored two bands in the DNA typing pattern, had failed to run appropriate controls, and had not applied its own quantitative criteria for matches. Lifecodes had thus interpreted its evidence in what could charitably be called a creative fashion. These lapses called into question the entire enterprise of DNA forensics.

The expert witnesses called by both prosecution and defense took the unusual step of going outside the courtroom to confer among themselves, and they prepared a report for the judge. The judge ruled that the evidence could be introduced to exculpate the suspect, but not to corroborate his guilt. The court clearly indicated that DNA testing was, in theory, admissible as evidence to identify the suspect positively, but doubts about laboratory procedure and interpretation in the current case made it inadmissible for that purpose. Castro pled guilty, and while the status of DNA testing in the case was thus not crucial to its outcome, the exposure of some pitfalls in DNA forensics became a lasting residue.

In other cases, Lifecodes presented statistics suggesting that the chances of a match were one in several hundred million or in the billions. The claims were outrageous given the paucity of the population-genetic database, which was held as a proprietary secret. Beyond the insufficiency of the population genetic data, there was always a possibility that the laboratory inadvertently switched samples, or that the person had a twin and did not know it (DNA typing, unlike standard fingerprinting, could *not* distinguish identical twins). The likelihood of such errors was obviously much higher than the figures being quoted in court. The *Castro* case ushered in a debate about laboratory practices, consistent band-matching criteria, and standards for interpreting the statistics.

The courtroom conflict between prosecution and defense began to spill over into the scientific community. A relatively small number of human geneticists, particularly those who were knowledgeable about both DNA typing and population genetics, were called as expert witnesses in many cases, but they did not agree among themselves about how to interpret the tests. The center of the controversy was the degree and significance of population substructure.

If the accused person came from a population that often had a DNA-type profile unusual in other groups, and if this group had few or no members in the population database used for interpretation, then the result could be highly misleading. Suppose, for example, that the suspect was a Basque and that Basques had type Z very frequently but other groups did not. There might be hundreds of thousands of Basques with that type. The database would not reveal this fact because it would include few Basques and would lump them with Caucasians. In the total database, the Basque pattern would appear quite rare. Moreover, Basques might be expected to cluster in the same neighborhoods, say the one where a murder or rape took place. If DNA types did indeed vary among subpopulations, errant statistics could make it seem that a match was far more significant than it actually was. Another possible source of errant interpretation was if two traits were assumed to be independent, but were actually associated with each other. Nordics, for example, might often have blond hair and blue eyes, but if the probabilities of blue eyes and blond hair, both uncommon traits, were multiplied together, it would seem extremely unusual for individuals to have this combination. There were few data to assess how often this kind of bias was present for the new DNA markers, and so there was ample room for divisive scientific combat.

The scientific controversy spilled onto the pages of *Science*. Two groups of highly competent population geneticists took opposite sides on this question. One article cast doubt on how forensic tests were being employed and asserted a need for considerably more data about the frequency of DNA types among disparate populations before the technique should be used to decide the fates of the accused. A companion piece, commissioned by editor Daniel Koshland, doubted that population substructure was large enough to mislead juries and judges.

Those interpreting DNA forensics typically used different base statistics, depending on the race of the suspect. This practice was troubling from both social and technical points of view. It was inherently disturbing to use race overtly in criminal proceedings. Moreover, the "racial" categories corresponded only poorly to population-genetic knowledge. The Federal Bureau of Investigation used a category of "Hispanic," for example, but this could refer to a person from a Caribbean island, someone from Aztec, Inca, or Mayan extraction, or someone whose ancestors came from Spain and Portugal—a hopeless mishmash.

The task of sorting out the technical arcana fell to the NRC committee. The NRC report was caught in a crossfire between the FBI and prosecutors on one side and defense attorneys on the other. The battle was joined by geneticists. The intensely adversarial ethos of the courtroom seared the professional egos of many, as their motives were impugned, inconsistencies amplified, and characters flayed not only before the jury but also in the public media. A network of prosecutors and a countervailing network of defense lawyers resorted to tactics of intimidation and persistent annoyance vastly more aggressive and personal than the usual intellectual fencing within science. Some scientists, although not the most prominent scientific authorities, and not a few lawyers made a living on the introduction of DNA forensics. Most of the fights centered on how to interpret laboratory results.

The 1990 OTA report noted this controversy and called for population geneticists to formulate standards. The NRC committee, as an expert body, attempted to do just that, to find consensus about how to interpret the results of DNA forensic tests. As the report approached release, the *New York Times* broke a story that concluded the report would recommend a moratorium on DNA forensics until there were better standards. This forced the committee to schedule a press conference in great haste, to dispel the call for a moratorium recounted in the story.

The NRC report made a series of significant recommendations. It called for an independent expert body, outside the FBI, to monitor laboratory practices and to make recommendations on how DNA forensic testing should be performed. The committee's most significant contributions but also its greatest vulnerability, came from recommendations about how to handle the population-genetic analysis. The committee reviewed evidence that population substructure was probably not a major source of errant matches, but it allowed that "population substructure may exist." It acknowledged that existing statistical databases were insufficient to determine the extent of population substructure, and argued that "the solution, however, is not to bar DNA evidence, but to ensure that estimates of the probability that a match between a person's DNA and evidence DNA could occur by chance are appropriately conservative." The databases should be made better, and this could be done by directly measuring the extent of population substructure among ethnic groups.

The controversy over DNA forensics thus reinforced the need for better data about human population genetics. This had already been raised as an urgent priority among anthropologists and paleontologists who hope to use genetics to understand human origins and historical migratory patterns. The need for robust forensic databases gave the same data a decidedly practical twist, with lives hanging in the balance. Controversies about how best to interpret population-genetic data that had long been obscure and of only academic interest were suddenly directly relevant to the fates of suspected criminals, and to the pursuit of justice. Whether data would dispel the fractiousness of the population-genetics community was open to doubt, but if not, there was a long future for careers in expert testimony. The first step, and the best hope, was to collect empirical data on human populations by going out and sampling them.

A recommendation to stop using race-specific analyses was also a major advance. The committee suggested that the frequency of any given DNA marker pattern should be interpreted to the benefit of the suspect, under a "ceiling principle." The number used to assess the likelihood of a match should be taken from the population group with the highest frequency. This default assumption was a clever way to get around the troublesome process of determining racial origin. Instead of trying to decide which "race" the suspect came from and applying different statistics for each group, the suspect's racial background would be irrelevant, and the suspect would be given the benefit of the doubt for each marker tested. If prosecutors needed more statistical power, they could order more markers to be tested, possible in many but not all cases. This proposal would bring down the probabilities from the ludicrous range of one in millions or billions, but the technique would still generally be far more reliable than eyewitness identification or blood-protein tests. The committee suggested setting arbitrary conservative probabilities until data began to flow in from the empirical population surveys.

The committee thus cut the Gordian knot, hoping to preserve the admissibility of the evidence, to shore up the regulatory framework, and yet to interpret the evidence in a conservative and scientifically defensible manner. It was not clear, however, how judges would react. In what was purportedly a review of books about the genome project but proved more a platform to air his views, Harvard geneticist Richard Lewontin noted that the NRC report might not resolve the population-genetic controversy. Judges might focus on the report's ambiguity, calling for empirical data, rather than accept the committee's interim solution of ceilings and arbitrary marker frequencies. A cautious response would favor waiting for more data about the markers being used in specific populations. The courts predictably differed in how tightly they embraced DNA forensics. Some accepted DNA forensic evidence, while others awaited resolution of the population genetic issues. While the eventual acceptance of DNA forensics was not in doubt, the speed with which it would become routine was highly uncertain and appeared likely to differ markedly among jurisdictions.

The controversy refused to die, and even as some courts began to admit DNA forensic evidence more readily, another controversy broke out in *Science* with the publication of an article critical of the NRC report and a news feature on the same topic. This time, the NRC committee was lambasted for erring too far on the side of conservatism. A group of population geneticists cast doubt on the significance of population substructure for those markers being used in forensic work. They disagreed with the logic behind the ceiling principle and asserted that most data suggested that marker frequencies could generally be multiplied together—the procedure that produced such astoundingly large odds ratios. They pointed out the need for much larger samples of population groups than those suggested in the NRC report to get sufficient data; a survey of the proposed size would generate relatively unreliable and unduly high estimates of marker frequency because the number of individuals sampled would be low and the margin of error correspondingly high. The upshot was that courts were being misled into a too conservative stance on DNA forensics by the NRC committee, and that the empirical surveys intended to solidify the basis for interpretation would not be sufficiently robust to restore balance. The NRC report was thus being attacked from both sides. Some critics claimed it was unduly conservative, while others contended the NRC committee had too readily accepted DNA forensics. The NRC commenced a second DNA forensic study in the summer of 1993, hoping to finally quell the controversy.

The courtroom entry of a new genetic technique, derived from gene mapping efforts, was noisy and slow. The problems emerged only as specific cases provoked scrutiny of existing practices. Abstruse questions of population genetics, a mathematically complex and relatively small academic subspecialty, were suddenly exposed to the harsh realities of the criminal justice system. Science collided with an adversarial court tradition, and the result was five years of turmoil, entailing hundreds of hearings, thousands of hours, and millions of dollars. In the wisdom of hindsight, the sources of controversy could have been resolved by empirical research, standardization of methods, and conservative race-neutral interpretation. But like the field from which it arose, DNA forensic testing took a bumpier road to acceptance. . .

III. GENETICS AND INDUSTRY

EDITOR'S INTRODUCTION

The new technologies that have been and are being developed for diagnosing and treating genetic diseases have not escaped the notice of venture capitalists and entrepreneurially minded scientists. A number of new biotech companies have been established in recent years, many of them with the aim of commercializing therapeutic treatments to genetic diseases. An overview of this development is provided in the first article, by Gene Bylinsky and reprinted from *Fortune*. Other biotech companies are hoping to profit from research being done on transgenic animals. The second article, "The Organ Factory of the Future?" focuses on the work of a British company that has created genetically engineered pigs whose organs, researchers hope, will not be rejected when transplanted into human patients.

The food industry has also been transformed by scientists' ability to manipulate genes. The third article, reprinted from *Chemistry & Industry*, presents an overview of the myriad ways in which genetic engineering is used in the food industry. Genetic engineering has also turned into a reality the idea of increasing the yield of various crops. The impact that this development could have on food production in the Third World is discussed in the fourth article, "Can Biotech Put Bread on Third World Tables?"

While a majority of Americans agree that genetic engineering should be used to produce more food, many are concerned about the safety of genetically engineered food products as well as about the effect that the manipulation of genes will have on the environment. These and other issues still under debate are covered in the fifth article, by Pamela Weintraub in *Audubon*. This section concludes with "No Human Risks," a piece by Kevin L. Ropp that touches on the recent debate over the safety of milk produced by dairy cows that have been injected with a genetically engineered growth hormone designed to increase the level of milk production.

GENETICS: THE MONEY RUSH IS ON[1]

Behind the red brick walls of two unprepossessing buildings in a science park in Rockville, Maryland, 135 scientists and entrepreneurs are laying the groundwork for a new epoch in biology and medicine. Computer-assisted robots, the galley slaves of the 21st century, work around the clock in spotless, brightly lit bay, doing the researchers' bidding. The object: nothing less than to decipher and commercialize chemical sequences that make up human genes, the molecular arbiters of health, intelligence, and behavior.

The researchers, robots, and computers are all part of Human Genome Sciences (HGS), the largest and the most lavishly financed of about a dozen new companies racing to crack the human genetic code. Some analysts think HGS is so far ahead of the field that it may be the Microsoft of genetics in the making. William A. Haseltine, who took a leave from a professorship at the Harvard Medical School to serve as CEO, boasts that HGS and the research foundation it supports, The Institute for Genomic Research, will have isolated and partly deciphered most of the important human genes within two years. Developing medical applications could take years more, but simply having the knowledge will give HGS what Haseltine calls the "gotcha" advantage. HGS expects to profit mightily by licensing the secrets it uncovers to drug companies and exploiting them itself. It should be able to patent many of those discoveries.

So far, the most visible manifestation of the gotcha advantage is a chauffeur-driven limousine, parked in front of HGS headquarters, at Haseltine's beck and call. The company turned its first profit last year, $1.8 million on $22 million in revenues, and went public in the fall at $12 a share; at a price of $16.75 a share in early May, Haseltine's 8% stake was worth about $20 million. The public offering succeeded in part because HGS has the backing of drug giant SmithKline Beecham, which committed $125 million a year ago for product rights and a 7% stake. HGS's 1993 revenues consisted entirely of payment from that arrangement.

[1]Reprint of article by Gene Bylinsky from *Fortune* v 129 pp94–108 My 30 '94.

Seldom has a major corporation put up so much money at such an early stage for a technology being developed by someone else.

SmithKline Beecham isn't alone. In recent months, platoons of scouts from big drug manufacturers were seen sliding on Southern California mud and slipping on Boston-area ice as they sought out startups with names like Millennium Pharmaceuticals (Cambridge, Massachusetts) and Sequana Therapeutics (La Jolla, California). Among the most hotly pursued: Myriad Genetics (Salt Lake City), co-founded by Walter Gilbert, the Harvard Nobelist; and Darwin Molecular (Bothwell, Washington), co-founded by Leroy Hood, who invented a gene-decoding machine as central to the new industry as the cotton gin was to textiles. In March, Hoffmann-La Roche agreed to put more than $70 million into Millennium for the right to turn genetic data about obesity and adult-onset diabetes into pills. Venture capitalists say more such deals are imminent.

The big boys have also started a brisk trade in what might be called gene futures. Eli Lilly paid Myriad Genetics $2.8 million for rights to a gene implicated in breast cancer. Genentech, the $650-million-a-year biotech pioneer, invested $17 million in GenVec (Rockville, Maryland) and set up an affiliate called Genomyx. Baxter International, SmithKline Beecham, Hoffmann-La Roche, and other large companies are also rushing to build up inhouse gene-hunting capability. Says Kevin Kinsella, a veteran venture capitalist and founder and CEO of Sequana Therapeutics: "As a venture capitalist, I've started seven biotech companies since 1982, but I haven't seen anything like it. This is biotech's counterpart of the Oklahoma land rush."

What lures the pioneers is the promise of a bonanza that could make the Sooner land rush pale by comparison. An almost biblical event—the final decoding of the fundamental secrets of life—will pay off, participants expect, in novel and uniquely effective medicines. With genetic codes in hand, scientists will be able to design drugs to attack the causes of disease rather than the symptoms, as most medications do now. Afflictions that have crippled and killed for millenniums—some cancers, rheumatoid arthritis, heart disease—will become amenable to treatment. So will such miseries as migraine headaches and obesity. Even baldness may be treatable—Kinsella's company hopes to develop a shampoo that would deliver genes for promoting hair growth directly to the cells of the scalp. Diagnosing people genetically prone to a given illness and treating them preemptively could reduce national health costs, perhaps by a lot. Exults Haseltine: "We are witnessing the beginning of a new age in biology and medicine."

The trip from the lab bench to the bank window is almost sure to turn out to be more difficult and time-consuming than scientists and investors expect. But for pharmaceutical companies, holding back could mean missing a historic opportunity. Juergen Drews, president of international R&D at Hoffmann-La Roche, hints why when he says that the new exploration of genetics "will revolutionize our approach to drug discovery."

Scientists will avoid many of the blind alleys of conventional R&D, still largely a hit-or-miss process that involves concocting new compounds and then testing them for therapeutic effects. Instead, researchers will be able to precisely design drugs aimed at specific targets. The promise of dozens of potent new medications comes at a time when the R&D pipelines of most big companies are almost empty. In the past three years, according to the Food and Drug Administration, big companies won approval for only 81 truly new drugs, including treatments for high blood pressure, epilepsy, and infections. The rest—998 medications—were variations on existing drugs.

The man largely responsible for this historic upheaval is J. Craig Venter, 47, director of HGS's nonprofit affiliate, The Institute for Genomic Research (TIGR) in Gaithersburg, Maryland. A soft-spoken molecular biologist, Venter presides over the most advanced gene-discovery laboratory in the world. In a cavernous room, fresh-faced young people in white lab coats poke at computer keyboards and ease tiny glass tubes into the maws of big, gray gene-reading machines. The vials are filled with molecules of human DNA and four fluorescent dyes, each designed to mark one of the four basic chemicals that pair up to form rungs of the famous double helix. Lasers activate the dyes and generate vivid, stained glass–like images on computer screens. The images correspond to the sequence of so-called base pairs in the DNA being tested.

Data from the machines course through thick black cables into a Maspar supercomputer, equal in power to 4,000 PCs, on the floor above. It compares the new codes with millions of genetic sequences from dozens of species. Because the genetic code is universal, a human sample may match DNA from another human, a rat, a bat, a mouse, a worm, a fruit fly, or even a microbe.

The human genome—the entire DNA code for a human being—differs from that of a chimpanzee in only about 1.5% of its content; chimps, fruit flies, and other creatures make proteins identical to or similar to those of humans. If the new sequence matches a known one, which currently happens about half the time, researchers can infer its genetic function; otherwise they carefully catalogue the sequence for further investigation.

Venter's procedures help scientist-entrepreneurs sidestep the biggest obstacle in biogenetics: the sheer vastness of the genome. The human body consists of more than 75 trillion cells, each of which, except for red blood cells, has a full complement of chromosomes nestled in its nucleus. Each chromosome is a wadded-up strand of DNA made of hundreds of millions of base pairs; stretched out straight, it would measure anywhere from three to nine feet long and about 20 atoms across. The chromosomes—46 to a cell, in 23 pairs—constitute a complete set of instructions for the making and functioning of a human being.

Those instructions take the form of genes, submicroscopic chemical sequences scattered along the chromosomes. Magic words in the book of life, genes direct the production of proteins that make up cells' structure and run their vital chemistry. Of all the DNA in the genome, genes account for scarcely 3%; the other 97%, sometimes called junk DNA, is now thought to serve structural and other purposes. The Human Genome Project, a $3 billion international effort to map the entire genome that was launched in 1990, involves 350 labs and isn't expected to complete its task until 2005. Since genes are distributed randomly, and sometimes in pieces, among the six billion base pairs in each set of chromosomes, isolating them represents one of biology's great challenges.

Working in the 1980s at the National Institutes of Health, Venter brilliantly combined three insights. He found a way to harness living cells to isolate genes. "A cell can do much more than any supercomputer," he explains. "It knows how to extract the information it needs from the chromosomes." Inside the cell, a remarkable mechanism transcribes DNA into a concise blueprint for a protein, called messenger RNA, by editing out all the junk DNA. The cell works much like an insomniac movie buff taping a film from late-night TV and zapping the interminable commercials. Venter decided to fish out the fragile messenger RNA molecules from cells, and by applying special enzymes, to transcribe them back into sturdy DNA that represented pure, edited genes. This so-called complementary DNA could be analyzed and held for future reference in vials in a freezer.

The discovery helped open the new frontier. Government agencies, universities, and private institutions have used Venter's method to build entire biological libraries. They keep samples of healthy and defective genes and make the deciphered codes available via Internet to researchers all over the world. Scientists use the data to investigate genetic defects, even the tiniest of which can cause devastating illness. For example, sickle cell anemia, the crippling blood disorder, results from a single error in a single gene. Other diseases, such as asthma and diabetes, are polygenic —associated with multiple genetic defects, often on more than one chromosome, interacting in ways not yet understood.

Venter was among the first to see that by marrying computers and biotechnical instruments, scientists could speed the search for the genetic causes of disease. He championed the use of robots to perform routine experiments in the lab; by hooking together machines that read genes and computers to process the resulting data, he was able to streamline the isolating and decoding process. The field he helped pioneer is now known as bioinformatics.

His third, most controversial, breakthrough accelerated the gene hunt by as much as a thousandfold. Rather than struggle to piece together entire genes, he decided to isolate and investigate gene fragments. Some scientists scoffed, arguing that analyzing fragments, especially those of genes whose function was not known, would be meaningless. But Venter was betting that the fragments would yield valuable clues to the meaning of other genetic data accumulating in computerized libraries, just as a scrap of a secret message can be used to crack a much longer message written in the same code.

Soon Venter was decoding DNA fragments by the thousand. His work exploded onto the commercial scene in 1991, when NIH attempted to patent the fragments as a public service. Bernadine Healy, the NIH director, intended to encourage commercial applications of Venter's research by licensing the data at nominal cost to private companies. She argued that companies would not be interested in developing gene-based medications unless they could be assured of patent protection. But James Watson, the legendary co-discoverer of the double helix and a director of the Human Genome Project at NIH, blasted the plan, arguing that patenting the secrets of life would hinder research.

Venter soon found himself in bureaucratic Siberia. When he applied to the Genome Project for funding to expand his lab, he was rejected twice; prominent university geneticists, such as David Botstein of Stanford, criticized his work as narrow and shortsighted.

The hubbub set off sensitive seismographs in the offices of biotech venture capitalists. Before long, Venter struck an innovative deal with Wallace H. Steinberg, a former Johnson & Johnson executive who heads HealthCare Management Investment Corp., a venture fund in Edison, New Jersey. They founded Venter's institute and HGS in 1992. In exchange for exclusive rights to the institute's discoveries, HGS is obliged to pay $85 million over a ten-year period. Venter and his 70 researchers are free to publish their findings after HGS and SmithKline Beecham have mulled them over for at least three months. Though he wound up with a stake in HGS that is now worth $12 million, Venter says he wasn't looking for money. What makes him happy is the freedom to pursue whichever genetic research he wants. Says Venter: "Now I'm the bureaucracy."

Venter's research was vindicated in March when a team from the Johns Hopkins Oncology Center in Baltimore used his institute and HGS to help pinpoint genes on Chromosome 3 that play a role in cancer of the colon. HGS, which has invested heavily in finding and cataloguing defective genes, had already singled out those in question. The colon cancer genes are particularly important because they hint at a mechanism that may be at work in many forms of cancer. In their healthy state they help direct the repair of damaged DNA within the body's cells. When they themselves are damaged, they no longer function. The defective DNA accumulates, and cancer soon starts to grow.

HGS and TIGR are racing to refine their gene-hunting technology and expand their competitive edge. But the work Venter did at NIH remains in the public domain, and many other genomics companies are using it too. HGS's biggest competitor, Incyte Pharmaceuticals in Palo Alto, is compiling data on genetic differences between healthy and diseased immune-system cells. The information could help companies design drugs against allergies, asthma, rheumatoid arthritis, and other immune-related disorders. Incyte's strategy, says director of business development Lisa Peterson, is to market its data to major drug companies: "Since HGS is largely tied up with SmithKline Beecham, we plan to serve the rest of the industry." Like HGS, Incyte went public last fall.

Other genomics companies are betting on an approach that is the reverse of Venter's. Rather than match gene fragments with diseases, they are working backward, trying to trace inherited disease to its genetic cause. Millennium, for example, has tackled three tough polygenic disorders: obesity, diabetes, and asthma. Myriad Genetics is probing the causes of heart disease and cancer by analyzing the vast records that the Mormons have kept of generations of faithful in Utah. Sequana hopes to solve the mystery of baldness by comparing human genes with those of hairless mice; it is trying to isolate obesity genes by studying mice that naturally get fat. Such quests will go on for many years. But the knowledge genomics researchers are accumulating may pay off in a surprising variety of ways, some quite immediate:

Diagnostics

Detecting disease is the arena in which many genomic start-ups are looking for their first commercial success. Analysts see a $7-billion-a-year industry taking shape in the next few years. One leader, Genica Pharmaceuticals of Worcester, Massachusetts, sold $7.2 million of tests for neurological disorders last year. With nearly 30 diagnostic tests on the market or under development, the company claims to have commercialized more discoveries stemming from the Human Genome Project than anyone else.

The promise of genetic diagnostics lies in the certainty with which such tests can detect disease. Matritech, a publicly traded company in Cambridge, Massachusetts, for instance, is developing a foolproof test for colon cancer. It checks for six proteins that turn abnormal only when cancer is present. Genetic tests may also help doctors gauge the severity of a disease, which can vary depending on which part of a gene is defective. For example, in cystic fibrosis, a congenital disorder affecting the lining of the lungs and pancreas, the gene in question on Chromosome 7 can be damaged at more than 300 sites. The severity of the disease depends on the locations of the flaws. Tests that pinpoint some of them are now being sold by Integrated Genetics of Framingham, Massachusetts.

The ultimate diagnostic tool may be the laboratory on a chip: tiny devices that combine a chemical process and electronic sensors to test for a specific disease. Such a product is under development at Lawrence Livermore National Laboratory in California. Support for the project comes from Roche Molecular Systems, a

Hoffmann-La Roche subsidiary in Alameda. The chip is a square of silicon less than a half-inch across; it incorporates a reaction chamber in which DNA fragments are heated and exposed to reagents to reveal the genetic code. The hope is to use the chip in a hand-held, battery-powered kit for diagnosing diseases in hospitalized patients.

Meanwhile, Affymatrix in Santa Clara, California, is working on a dime-size lab-on-a-chip. It is designed to compare a DNA specimen squirted onto its surface with embedded test samples of DNA. Defects show up as mismatches, which chemicals on the chip cause to fluoresce. A laser device scans the chip and displays the result on a computer screen.

Gene Therapy

Diseases such as muscular dystrophy, hemophilia, and some forms of heart disease and cancer are characterized by the body's inability to manufacture a vital protein. Soon doctors may be able to administer new genes that restore protein production. Early attempts at such treatments have worked. W. French Anderson, a former NIH researcher now at the University of Southern California, has healed a small number of children with rare genetic diseases by injecting them with undamaged genes.

Preliminary trials on patients with cancer, cystic fibrosis, and other diseases are already under way. The question of how best to administer replacement genes is one of the most intriguing puzzles in medicine. Ideally, defective genes should be detected and replaced in childhood or even in the womb, before they cause disease or get passed along to future generations. But technology is nowhere near that stage, and researchers are pressing ahead with the tools they have.

To treat cystic fibrosis, GenVec of Rockville, Maryland, is testing a liquid it hopes to use in spray form to deliver healthy DNA to the lungs. In March, researchers at the University of Pennsylvania Medical School reported that they had introduced healthy genes surgically into the liver of a young woman from Quebec who suffers from an extremely high cholesterol level. The researchers removed tissue from her liver, introduced intact genes into the cells, and then reinserted them. Her cholesterol quickly dropped 20% as the new genes began to function; it is too early to tell whether the improvement is permanent. Genetic Therapy, of Gaithersburg, Maryland, is among the companies positioning themselves to commercialize such research.

Regulating Genes

A gene that fails to function doesn't necessarily have to be replaced. Sometimes all that is defective is its on/off switch—the sequence of DNA that starts or stops the protein production process. Replacing command sequences, which are as short as 50 base pairs and are readily synthesized, could represent an entirely new way to treat disease. Transkaryotic Therapies of Cambridge, Massachusetts, recently showed in the lab that it can insert a new switch into a cell and turn on a balky gene. In the future, expensive manufacturing plants may no longer be needed to produce proteins like erythropoeitin (EPO), which restores red blood cells in anemia; the patient's own cells may be stimulated to do the job.

Protein Therapy

Genome companies also hope to use their burgeoning knowledge to produce medicinal proteins in the lab, using well-established genetic engineering techniques. Haseltine notes that biotech powers like Amgen and Genentech have built sizable businesses on just a few such drugs. Example: Genentech's tPA (tissue plasminogen activator), which dissolves blood clots in heart arteries. "We already have 200 to 300 genes for candidate proteins," says Haseltine. Because HGS and other genome startups are allied with major drug manufacturers, new proteins could find their way to market within the next few years.

Small Molecule Drugs

Test tube chemistry can be harnessed to make small molecules for the treatment of many diseases including AIDS and some cancers. Unlike proteins, which are effective only when injected directly into the bloodstream, small molecules pass easily through the lining of the stomach and can be made into pills. That's where the big money is in the drug industry.

Darwin Molecular of Bothwell, Washington, is staking its future on a daring strategy to convert genetic sequences directly into candidates for pills. In a novel approach called man-made molecular evolution, it uses chemistry and computers to design and test millions of molecules for their ability to find and repair disease-causing genetic defects. In a series of screening steps, the candidates are gradually weeded out; only the fittest make it through the entire process and go on to clinical tests. So promising is the method that Darwin recently attracted a $10 million investment from Microsoft billionaires Paul Allen and Bill Gates.

Like dirt farmers who became rich in the land rush by discovering oil on their property, some genomics entrepreneurs will likely hit the jackpot in genes in the next decade. Says Sequana's Kinsella with total conviction: "The payoff of genomics will be bigger than that of the Manhattan Project or the space program." But there are epic risks as well: Many startups will fail as founder-researchers prove to be less adept at management than at science, and giant companies that move too slowly may see their markets eaten away.

To fully exploit the promise of genomics will require effort on a scale that is hard to comprehend. James Watson of double-helix fame predicts that it will take geneticists 10,000 years to fully understand the workings of the genome. Still lacking, for example, is an explanation for how genes dictate the three-dimensional shape of proteins. Within cells, proteins act like biological micromachines. Some form valves that let nutrients enter through cell membranes; others serve as catalysts in chemical reactions. Yet no scientific language today can comprehensively relate this 3-D reality to two-dimensional codes of DNA. That is only one of the untamed frontiers that still confront genomics pioneers.

THE ORGAN FACTORY OF THE FUTURE?[2]

At a secret location in Cambridgeshire, researchers inject human DNA into a pig embryo. Six months later Astrid, the world's first transgenic pig, is born—of a virgin, in a sterile stable, on Christmas eve. The hope is that the implanted gene will make pig organs compatible with the human immune system, thus helping to solve one of medicine's fastest growing problems: the shortage of organs for transplant surgery. Astrid produces offspring, the research gathers pace. But there are problems, too: antivivisectionists launch firebomb attacks and medical ethicists get jumpy.

It could be the plot of a TV drama about the future of genetic

[2]Reprint of article by David Concar from *New Scientist* v 142 pp24–9 Je 18 '94.

engineering, but it isn't. Astrid, the "pig with a human heart" as she is dubbed in headlines, is as real as the surgical aspirations of the British scientists who created her two years ago. Now the transgenic clan has grown to some 200 pigs. And Imutran, the company behind the project, is taking the next step—testing what happens when human blood is pumped through hearts taken from some of Astrid's descendants.

This week, Imutran's research director, David White reports the first findings to delegates at an international conference on "xenotransplantation" in Washington DC. "It's absolutely clear that hearts from transgenic pigs work better than those from normal pigs and show fewer signs of immune rejection," he told *New Scientist* beforehand. "But what isn't clear is whether that result will correspond to better survival rates of xenografts in primates or humans."

If it does, the rewards—financial as well as medical—could be considerable. Last year in the US alone some 2800 people died waiting for human organs to become available, and in Britain at present about 25 per cent of heart patients die waiting. Such figures are set to climb next century as conventional transplantation techniques improve and patients who are now considered too old or sick to benefit from new organs are put on waiting lists.

All being well, Imutran expects trials in humans to begin in 1996. But hard on its heels are two American biotechnology companies replete with glossy brochures featuring transgenic pigs, some designed to function as organ donors. Moreover, the American companies have already started transplanting genetically-altered pig tissue into primates—experiments whose outcome will be vital to persuading ethics committees that there is a case for proceeding to trials in humans.

Surgeon's Dream

A bountiful supply of designer organs sounds like a transplant surgeon's dream. But before it can be turned into reality, Imutran and its rivals must clear some towering obstacles. First, nobody has yet begun to draw up guidelines on the ethics of using genetically-engineered animal organs in surgery, and especially how to preempt any needless experimentation on humans. Falter here, and the pig engineers could find the public set against them. Secondly, the law on patenting transgenic animals is still desperately unclear, raising the prospect of a legal dispute over commercial rights. But the biggest question of all rests with biology: can genetic engineering really deliver animal organs that look like "friend" rather than "foe" to the human immune system?

At the moment, everyone is focusing on making pigs with genes designed to disarm a powerful immune response known as hyperacute rejection. After conventional transplants, organ rejection can, in most cases, be prevented with drugs—such as cyclosporin—that act to "handcuff" aggressive white blood cells known as T cells. But when organs are transplanted from species to species, rejection is too fast and violent to be pacified with drugs alone. The immune system treats the graft much as it would a solid clump of bacteria, unleashing agents that destroy the epithelial cells at the surface of the grafted organ while triggering a massive clogging of the arteries supplying it with blood. Within hours, the graft is reduced to a blackened mess.

Much of the damage is caused by a team of hostile blood proteins called the complement cascade. If antibodies and white blood cells are the ground troops of the immune system, the complement cascade is its air force. Normally reserved for attacking microorganisms, its proteins can punch lethal holes in cell membranes, producing an effect which, in White's words, is "like a bomb going off".

It was a desire to defuse this bomb that led the Cambridge researchers to inject human DNA into pig embryos in August 1992. Human cells are spared from attack because they carry markers—molecular "white flags"—that can pacify complement proteins. The researchers reasoned that if they could transfer genes for these white flags into pigs, the complement cascade might be fooled into holding its fire. One candidate for the job was a gene encoding a protein called Decay Accelerating Factor. In human tissue, it was clear that DAF molecules stuck out of the surfaces of cells, warding off complement proteins. Could they do the same in pig tissue?

The latest results on perfusing transgenic hearts with human blood, reported this week in Washington DC, suggest the answer is a qualified "yes". "We see little or no signs of any hyperacute rejection of pig hearts expressing DAF," says White, who nonetheless stresses the limitations of perfusion tests as a measure of immune compatibility.

Mouse Hearts

A similar picture is emerging from Imutran's rivals in the US. About 18 months ago, Boston-based DNX Corporation began producing mice and pigs carrying not only DAF but human genes encoding two other "white flag" proteins, CD46 and CD59. Experiments involving perfusion with human blood show that transgenic mouse hearts are not attacked by the complement cascade, says John Logan, DNX's vice-president of research.

The latest company to join the race, a Yale firm called Alexion Pharmaceuticals, is hoping to gain ground with a genetically-engineered protein designed to combine the talents of both DAF and CD59. Having just filed for a patent on the protein, Alexion is cagey about details but decidedly upbeat about clinical potential. "We've taken engineered pig epithelial cells that produce high levels of protein and transplanted them into primates," says Stephen Squinto, the company's programme director. "Normal pig cells are destroyed by hyperacute rejection within minutes. The engineered cells survived for hours."

Alexion expects to be transplanting designer pig organs into primates this autumn. "If we can get decent survival, we'll move on to humans," says Squinto. "We'll probably start with high risk patients and will most likely transplant hearts. The heart is a priority because there's little you can do with dying patients. You can't put them on dialysis."

If the Cambridge researchers are anxious about this competition, they are certainly not showing it. "Everything is beginning to hot up," says White, "but we like to think we're still in the lead."

Even those who normally preach caution on xenotransplants seem to be caught up in the excitement. Mindful of past failures to transplant baboon organs into humans, Roy Calne, a pioneer of kidney transplantation and a surgeon at Addenbrooke's Hospital in Cambridge, warned researchers at a conference in Cambridge last autumn against rushing into more xenotransplants on humans without getting a firmer grip on the biology that causes such transplants to be violently rejected. However, in the next breath, and only half in jest, Calne conjured up a new era in transplant surgery based on the "self-pig".

One day, the vision goes, transgenic technology may be so cheap and easy that we may all take the precaution of paying for the creation and upkeep of a custom-made transgenic pig, an immunological twin in porcine clothing that would come to the rescue in the event of an accident or disease. Contract hepatitis, and self-pig would provide a new liver; develop Alzheimer's disease, and a supply of personalised pig neurons would be at hand. Heart failure? No problem.

It sounds far fetched, and for the time being it is. Yet a decade from now, self-pigs may fall within the reach of genetic engineers. But whether such animals will ever see the light of a sanitised sty is another matter.

The genes that enable our immune systems to distinguish "self" tissue from foreign tissue, and which make each of us immunologically unique, are encoded by a vast tract of DNA known as the "major histocompatibility complex". To make self-pigs, you would have to disable each animal's MHC genes and replace them with copies of those belonging to each human "twin". But until recently that would have been unthinkable on technical grounds. Gene "knock out" techniques were too laborious and imprecise, and it was possible to transfer only relatively short pieces of DNA from animal to animal.

Now things are quietly changing. Researchers bent on reprogramming animal genomes are discovering the benefits of "yeast artificial chromosomes", which can be used to transfer stretches of DNA as long as 500,000 base pairs or more. And that could revolutionise the whole business of making transgenic pigs.

Take the case of Astrid and her siblings. White and his colleagues created them by injecting an embryo with an artificially "edited down" version of the natural DAF gene. Partly because of its size the gene is expressed somewhat erratically: not all organs in all pigs bear the DAF white flag. If nothing else, say the researchers, YACs could help to solve that problem by allowing the insertion of a much fuller version of the gene.

But scientific feasibility is only part of the equation. Just as important is commercial viability. And this is where the idea of self-pig could fall into the trough. To prospective backers, donor animals that must be genetically tailored for each and every patient years in advance of any medical problem would surely seem more like a legal and financial nightmare than a life-saving innovation. Or as Squinto puts it: "I don't see how you could market such animals."

Nor, sadly for fans of self-pig, does there seem to be any middle ground between creating "generic" organ donors—animals that could be used by everyone—and creating animals with genes specific to individual patients.

Yet commercial promise alone will not be enough to speed generic pig donors from the laboratory to the clinic. For a start, militant antivivisectionists in Britain are unlikely to call off their campaign of threats. Even moderate animal rights groups will continue to lobby for a European moratorium on the genetic manipulation of animals, or at least for restrictions on the patenting of such animals. In the case of transgenic pigs, they fear that tampering with immune genes could harm the animal, perhaps by causing immune disorders or a loss of resistance to infections that could be inherited.

Others also see this kind of genetic engineering as something of a slippery slope. "You're not treating the animal as an end in itself but as a means to an end," says Richard Nicholson, editor of the *Bulletin of Medical Ethics*. And has anyone bothered to consider the psychological impact on patients of using animal organs in transplant surgery?

Xenograft researchers react to such concerns with the air of an elephant staring down the barrel of a peashooter. There is no evidence of ill-effects in any of the transgenic pigs and mice produced so far, insist all three companies. And, as if to outface the worriers, conferences on xenografting seldom run seminars on ethics. Instead, one view on animal rights is invariably changed like a mantra from the podium: the idea of using pigs as organ donors is on a moral par with eating bacon. Speakers concede that primate donors, with their social hierarchies and seemingly richer emotional lives, might never be acceptable to the public. But who could question using an animal that is bred by the million for food and whose heart valves are already being inserted into humans?

Playing God

This line on ethics (by strange coincidence) harmonises perfectly with the practicalities. Baboons are slow breeders and are difficult to keep free from viral infections, some of them potentially lethal to humans. Pigs, by contrast, are about the same size as humans and can more easily be bred in sterile conditions. And surely the prospect of saving thousands of human lives justifies slotting the odd human gene into a porcine chromosome?

Perhaps. Yet even if proponents of xenografting triumph over animal rights (as seems likely), that won't be the last of the social obstacles. Just as worrying for the public is the surgeon-playing-God scenario, the fear that transplant teams armed with genetically-engineered animal organs will indulge in reckless experiments on human patients.

In the past decade in the US, there have been three attempts to transplant baboon organs into human patients, all failing, and all generating storms of controversy. Despite that, international guidelines on xenotransplants are still nowhere in sight, and decisions about operations remain in the hands of individual hospital or regional ethics committees. National medical guidelines, it is true, stress the need for "reasonably informed consent" in all medical experiments. But what this would amount to in the case of a xenograft experiment is far from clear.

At last year's conference in Cambridge, delegates were shocked to hear that surgeons at Cedar-Sinai Hospital in Los Angeles had already attempted to transplant a pig's liver into a 26-year-old woman. Within a day the organ had been rejected. "What you saw in this experiment was what you would predict from 30 years of literature—blockage of arteries and a mass of hyperacute rejection," says Squinto. "Without agents to block the hyperacute response, the experiment was premature." The condemnation seems unanimous. "Before doing this kind of trial on humans you have to show that the grafts can survive in animals," says Logan. "But that wasn't the case."

Will it be the case when surgeons want to experiment with transgenic pig organs? Might not the intense commercial competition encourage recklessness? Imutran and its rivals insist that a premature transplant with a negative outcome is the last thing they want. "We wouldn't want to go into clinical trials until we can be sure of getting survival rates similar to those for human-to-human transplants—a 70 to 75 per cent chance of surviving for more than a year," says White. Any trade in, and clinical use of, transgenic organs could be adequately policed by government watchdogs such as the Food and Drug Administration in the US, argues Logan.

A plentiful supply of transgenic animal organs might even help to reduce some long-running ethical concerns about transplantation surgery, say some researchers. The current shortage of human organs requires surgeons to make tough decisions about who should be put on waiting lists. Should organs go to the sickest patients, or those who can most benefit from them? And what about age? In Britain, there are no official age limits, but the vast majority of heart recipients are under 60.

Animal organs could change that. And in the long run, say their proponents, they could also lead to patients being spared some of the unpleasant and debilitating side-effects of immune suppression. "Xenotransplantation swings the focus away from interfering with the host immune system to interfering with the organ," says Squinto. "Why use drugs to produce broad suppression of the host's immune system when you can modify the graft?"

Even so, reassuring the public may require an openness about data that conflicts with commercial ambitions. Eager to protect the interests of their shareholders and backers, all three companies breeding transgenic pigs have filed for patents. As night follows day, an unseemly courtroom battle is now on the cards and a veil of secrecy hangs over many experimental details. For example, DNX declined to explain its gene constructs and experiments on primates two weeks ago citing "commercial sensitivity".

To pursue their commercial rights, the American companies may have to challenge a patent filed by the Cambridge researchers in 1989 which embraces the whole concept of using genes and proteins to protect animal tissue from attack by human complement. There are broader problems looming, too. For lawyers, the awkward thing about pigs is that they procreate. If company A makes a transgenic animal using techniques owned by company B, then it is certainly infringing the patent. But is that still the case if company A breeds offspring from this animal and sells them? And what happens if company A tries to sell a kidney from these offspring? "The issue is at best cloudy," says White.

Big though these social and commercial obstacles may be, they might ultimately be dwarfed by a more fundamental question: can animal organs ever be refashioned sufficiently to be fully compatible with the human body? Certainly, say die-hard optimists such as Stephen Grundy, a surgeon at the Medical Center of Loma Linda University, where the first baboon-to-human transplant was carried out a decade ago: "There are no biological barriers to xenotransplantation, just a series of small steps."

Others are more measured in their analysis. It is not just a question of immunology, says Calne. In addition to looking like a "friend" to the immune system, a transplanted organ must function properly too. "Even proteins produced by close species such as the baboon are different from their human counterparts," he notes despondently.

Even when it comes to immunology, the science of xenografting is still desperately young. "Five years ago it had the status of alchemy," quips one surgeon. It could take researchers years to hit on exactly the right combination of drugs, antibodies and transgenic donors to reach their final goal of producing complete tolerance.

A key worry is whether blocking the complement cascade will

be enough to prevent hyperacute rejection of pig organs in humans. The complement cascade is certainly important: animals born with genetic defects in complement genes are unusually tolerant to xenografts. But it might prove to be just the first of many hurdles. There is increasing evidence, for instance, that antibodies also contribute to hyperacute rejection. Everyone seems to carry antibodies in their blood that can attack pig tissue, and over the past two years it has become clear that a main target of these antibodies is a sugar molecule called gal(alpha-1,3)gal which is found on the surfaces of pig epithelial cells.

Cells of Wrath

Removing this sugary target could be vital to eliminating the hyperacute response, says Squinto. One approach would be to disable, or "knock out", the pig gene that encodes one of the enzymes needed to tether the sugar to the surfaces of cells. Making transgenic pigs of this kind is no easy task as it requires special manipulations of embryonic stem cells. But with Alexion and others now poised to try, there seems little doubt that "knock-out" pigs will be among the donor animals of choice next century.

And if they are, genetic engineers won't necessarily stop at removing the pig's troublesome sugar molecule. For even if the threat of hyperacute rejection can be completely silenced that way, xenografts will still be subjected to the gentler wrath of T cells—just as human-to-human transplants are. In conventional surgery, hostile responses of T cells are suppressed with drugs. But with pig organs, those responses could turn out to be much stronger. In which case, say researchers, it may be necessary to identify exactly which pig molecules trigger responses in human T cells, and eliminate them with genetic engineering. Commenting on the work still to be done, White says: "At the moment we're making a Model T Ford. But everyone would like to make a Ferrari."

But in the end, even a Ferrari-style donor may not be quite enough. To produce tolerance to animal organs, it may still be necessary to manipulate the patient's immune system; to administer "friendly" monoclonal antibodies that bind to antibodies that would otherwise attack pig tissue; even to transplant bone marrow tissue from a pig to the patient.

Bone marrow cells being the progenitors of all the body's immune cells, a pig-to-human transplant of this kind could—in theory—produce a host immune system with a conveniently split personality: enough donor immune cells to induce specific tolerance to donor tissue; and enough host immune cells to sustain normal human immunity. Some see this as a recipe for chaos in the immune system. But this week in Washington DC Elliot Lebowitz and his colleagues at BioTransplant in Charlestown, Massachusetts, report what they claim are promising results on rodents and primates.

In one experiment, the researchers used the bone marrow approach to transplant kidneys between primates of different blood groups, an operation that can be as risky as xenografting. The animals have survived for over a year, says Lebowitz—and without needing to have their immune system suppressed with drugs. The key to success, he says, is to kill off "mature" T cells in the recipient and remove troublesome antibodies before injecting the bone marrow cells and transplanting the donor tissue. That way, he adds, "a new immune system can be generated that accepts both the donor tissue and self tissue".

Lebowitz is convinced the dream of producing tolerance to xenografts can become a reality. He is not alone. "It's time again for xenotransplantation in the clinic," said Grundy uncompromisingly in Cambridge last autumn. An evangelist for the cause, Grundy likes to confront sceptics with a slide show of transplant heroes: a group of "xeno-goats", alive and well despite their sheep hearts, and a grinning baboon called Max who survived 502 days with a heart from a rhesus monkey and a little help from immunosuppressive drugs. How long before this gallery includes the face of a human being, alive and well and living with a pig's heart?

GENETIC ENGINEERING IN FOOD BIOTECHNOLOGY[3]

Gene technology is about to change the food market, and diverse fields for the application of recombinant DNA techniques are emerging. Food ingredients and enzymes for food processing

[3]Reprint of article by Heike Dornenburg & Christine Lang-Hinrichs from *Chemistry & Industry* 13, pp506–10 Jl 4 '94. Copyright © 1994 by Society of Chemical Industry. Reprinted with permission.

are produced by fermentation techniques using genetically modified microorganisms. Transgenic plants and animals for use in food are bred using modern genetic methods.

In general, the term '*genetic engineering*' describes a set of methods designed to change the genetic information of a cell or an organism directly. It provides the means by which genetic information (DNA) can be transferred across species barriers that cannot cross by conventional breeding. This can mean that: (i) a gene which has formerly not been part of the host's genetic information is added to a cell; or (ii) a gene is inactivated so that it can no longer synthesise a protein; or (iii) a gene is modified to synthesise a protein of an altered specificity and composition, or signal regions are manipulated so that higher yields or different regulation can be obtained.

In some cases the modified organism itself is part of the food product, such as a genetically engineered crop plant. In other cases an enzyme produced by genetically modified organisms or a starter culture is used to process the food and is removed before consumption.

Food Ingredients

The number of food ingredients that can be produced by genetically modified organisms has increased immensely over the past few years and now includes amino acids, vitamins, small peptides, flavours, and food enzymes.

Amino acids are needed as bulk substances for incorporation into food and feed. Most of the amino acids are made using strains of microorganisms developed by conventional genetic techniques. There are, however, some examples of processes involving recombinant DNA. In these cases the microbial metabolism has been manipulated to ensure the overproduction of one amino acid. The production of tryptophan by *Escherichia coli* is one such example. Recombinant strains contain either a deregulated version of a key enzyme for which precursor substances are supplied in the medium, or a deregulated operon that contains all the genes of the pathway is inserted in multiple copies. Other amino acids synthesised by recombinant strains are lysine, phenylalanine, threonine, and arginine.

The only vitamin for which a recombinant DNA process has been established is vitamin C, or ascorbic acid. A gene from the bacterium *Corynebacterium* has been introduced into *Erwinia her-*

bicola so that it can now use glucose as a direct precursor for 2-L-ketogluconic acid, a key metabolite of vitamin C synthesis.

A small peptide for use in food is the sweet thaumatin. It was originally a plant protein (isolated from the fruits of the African shrub *Thaumatococcus daniellii*) which has a sweetening power 2500 times that of sucrose. The gene encoding the protein has been cloned and expressed in the baker's yeast *Saccharomyces cerevisiae*. The yeast-derived product is identical to the plant product both from the biochemical and the functional point of view.

Another example is the sweetener aspartame which is a dipeptide of aspartic acid and phenylalanine. An *E. coli* strain has been constructed which produces a protein of repeated subunits of these amino acids. The protein can be isolated and enzymatically cleaved to yield aspartame.

The synthesis of flavours and pigments which are often plant-derived is also subject to genetic manipulation. In this respect the use of transformed plant cell cultures is most promising. Targets are plants such as *Beta vulgaris* for the synthesis of pigments (betacyanin, betaxanthin) or *Mentha piperita* for the synthesis of flavours (menthol, linalool, 1,8-cineol).

The production of enzymes for use in food and for the processing of food is one of the main fields of recombinant DNA technology in the food industry. There are already many examples of food enzymes whose synthesis has been subject to genetic engineering. . . .

Bulk amounts of enzymes are needed in the sugar industry to convert starch to glucose or fructose. As a result, several glucoamylase, α-amylase and glucose isomerase genes have been cloned and expressed in diverse host organisms. Pectinolytic and proteolytic enzymes have been produced for the fruit industry for extracting and clarifying juice. α-Amylase has also been used in the baking industry for starch degradation; glucanases and xylanases have been used for degrading glutan present in flour. Enzymes can also help to preserve fresh products without the use of chemical preservatives. Chitinases act against fungi, glucanases against yeasts and lysozyme against bacteria.

Recombinant rennet or chymosin is one of the most advanced examples. Chymosin is a calf-derived enzyme used as a clotting agent in cheese production. Traditionally, an enzyme preparation from calf stomach is used specifically to cleave κ-casein and induce clotting. The chymosin gene has been expressed in different host cells, and recombinant calf-chymosin is now commercially available either from the bacterium *E. coli,* the yeast *Kluyveromyces lactis* or the fungus *Aspergillus awamori.* Enzyme preparations are marketed as *CHY-MAX* (Pfizer), *MAXIREN* (Gist-Brocades) and *CHYMOGEN* (Christian Hansen's Lab and Genencor), respectively. These products are already used in several European countries, including the UK, Switzerland, Italy and the US for cheese production.

The enzyme acetolactate decarboxylase (ALDC) has been cloned in *Bacillus subtilis* and is marketed under the brand name of *MATUREX* (Novo Nordisk) for application in the brewing industry, where it is used to prevent the formation of diacetyl and thus to reduce maturation time.

While these recombinant enzymes remain in the final food product, other enzymes used for food processing are removed before confectioning. A lipase from *Rhizomucor miehei* produced by *Aspergillus oryzae* is, for example, used in an immobilised form for the production of speciality fats and new fats with improved nutritional qualities.

Starter Cultures

Genetically modified organisms can be used as starter cultures in the brewing and baking industry as well as in dairy and meat products. While the production of food ingredients does not leave any cells or organisms in the final product, starter cultures may well become part of the product. . . .

A genetically modified baker's yeast strain which continually produces CO_2 in dough has been introduced into the market by Gist-Brocades. Normal yeast strains exhibit an adaptation time during fermentation when glucose has already been consumed in the dough and maltose has yet to be metabolised. The new strain however, exhibits an accelerated maltose fermentation which is no longer strictly regulated and starts while glucose is still present.

Brewing yeast is also subject to modifications by recombinant DNA techniques. There are three main points of interest: (i) strains which produce less diacetyl (which is an off-flavour produced during brewing and which has to be removed during the long maturation period at the end of fermentation); (ii) strains which can express glucoamylases and ferment the starch present in the wort and which can be used to produce low-calorie beers; and (iii) strains which synthesise glucanases and thus remove polymeric β-1,3-1,4-glucanes which cause filtration problems and may cause undesirable haze in beer.

Prokaryotic microorganisms such as lactobacilli and lactococci are used for fermenting dairy and meat products. Genetic vector systems have been developed for some of them which allow their manipulation. A strain of *Lactococcus* has recently been made to synthesise lysozyme which efficiently inhibits *Clostridium tyrobutyricum,* responsible for the late blowing of cheese.

Lactococcal starter cultures often exhibit a certain instability. This is because genes responsible for key characters reside on plasmids and are not integrated into the chromosome. Stable strains have been constructed by integrating genes for specific proteases. Protease activity has also been increased by changing the expression signals of the gene. Such strains should enhance cheese ripening. Phage sensitivity is an important problem in lactic acid starter cultures. Insensitive industrial strains have been created by transferring phage resistance genes. Production of bacteriocins which work against many pathogens and spoilage bacteria can also be introduced into starter culture strains.

Starter cultures for meat products (fermented sausages) are also strains of *Lactobacillus* species. Here again bacteriocin production and the stabilisation of production characters are the target traits for genetic manipulation.

Transgenic Plants

Recombinant DNA technology is used to generate improved plant varieties for crop agriculture. New genes can be introduced from widely different plant species, as well as from microorganisms or animals, and provide new sources of genetic diversification. This technology will not replace, but will increasingly complement existing plant-breeding techniques. Both techniques can be combined to produce a range of plants with desired characteristics.

Bacillus thuringiensis, a soil bacteria, produces crystalline proteins which are toxic to insect larvae. The toxicity is caused by protein fragments released by the action of mid-gut proteases and their binding to receptors on the epithelial membrane which subsequently loses its selective permeability. The genes coding for these proteins have been cloned, characterised and mobilised into a number of crop plants, which have thus acquired resistance to insects. *Bacillus thuringiensis* endotoxin is the most commonly introduced gene for pest resistance. Many strains of *Bacillus thuringiensis* with a family of proteins acting against different crop pests have been discovered. Several food plants have been genetically engineered to express one or more endotoxin genes from the bacteria and have advanced to the field-testing stage.

Plant diseases are caused by a range of microorganisms including viruses, viroids, bacteria, fungi and nematodes. The greatest progress in conferring resistance to pathogens has been based on viral coat protein genes. Resistance to fungal pathogens is conferred by specific plant genes or genes from other sources (for example, bacterial genes which produce pesticides or chitinases). Transgenic crops resistant to fungi are potato (late blight), various vegetables, and tobacco for *Rhizoctonia* disease. Introduction of resistance to *Rhizoctonia solani* by genetic modification of chitinase in potatoes is also under development.

Several crop species have been genetically engineered for herbicide tolerance and have reached the field-testing stage. Most crops in Europe are being engineered to tolerate glufosinate, a relatively benign compound. An enzyme which degrades the herbicide was cloned from the soil bacterium *Streptomyces*. This enzyme functions in sugar beet, potato, soybean, canola and corn. By using glufosinate-tolerant sugar beet, growers could potentially cut the number of herbicide applications from a maximum of six to a maximum of two.

Controlled ripening of tomato fruits was achieved by inhibiting the genes coding for polygalacturonase. Recombinant DNA techniques were used to introduce genes into a host plant that code for antisense RNA which is not expected to be translated into a functional protein. Polygalacturonase degrades pectin causing softening of the fruit. Introduction of the antisense sequence of this polygalacturonase gene results in decreased synthesis of the enzyme and therefore in a prolonged shelf life. In the Calgene report, food safety aspects of the newly developed tomato are considered systematically. The *Flavr Savr* tomato, which has been engineered for better ripening on the vine, is available all year round. It is now the first genetically modified whole food to be available to American consumers. Such genetic modification of tomatoes will permit shipping of vine-ripened tomatoes, improved quality and shelf life and improved food processing quality.

An alternative approach to engineering tomatoes for improved storage suppresses a step necessary for the initiation of ripening. Introduction of the ACC synthase gene into tomatoes resulted in the arrest of all major ripening responses including colour and flavour development, softening and increased respiration. Controlled ripening is also under development for pea and pepper as well as tropical fruits.

Work on many food products using antisense gene technology for enhanced nutritional value is in progress. For example, canola oil is being modified to provide texture and consistency to margarine without the cholesterol-raising *trans* fatty acids that result from hydrogenation. Other canola plants are being antisense engineered to produce oils for the cacao butter market. In oil crops, fatty acid composition was modified to increase stability at high temperatures, to increase the melting point, and to provide a temperate crop substitute for palm oil. There is evidence that the nutritional quality of some plants can be increased by changing the amino acid content of seeds. Coffee which is low in caffeine content is also under development.

Transgenic Animals

Transgenic animals have been genetically transformed by the interspecies transfer of 'foreign' genetic information. Some of the goals in generating transgenic animals are to enhance production traits, to confer parasitic or disease resistance, and to produce novel or modified proteins. Gene transfer technology is likely to play an important role in the food production industry and can be used either directly or indirectly to improve the efficiency and quality of food production from mammals, birds and fish.

Many studies of transgenic animals have been concerned with attempts to increase growth or the efficiency of growth. The most dramatic example of manipulating growth by gene transfer is that of mice, where overexpression of human growth hormone genes stimulated growth with a fourfold increase in growth rates producing mice twice the normal size. The strategy has been extended to commercially important species and to increase the growth rates of livestock, fish and poultry. In contrast to the results obtained with growth hormone-transgenic mice and the administration of growth hormone to economically important animals, the corresponding growth hormone-transgenic animals did not show the same biological effects. Transgenic pigs and rabbits expressing the growth hormone gene did not show an increase in growth rates, but rapidly growing or 'giant' fish can be engineered by the addition of recombinant genes for growth hormones. The genetic improvements in fish are increased production efficiency, increased rates of growth, disease resistance, and extended ecological ranges.

Transgenic strategies which increase reproductive output, for example by increasing litter size or egg-laying capacity in poultry, can also influence the efficiency of food production. While the introduction of prolificacy genes to livestock with low ovulation rates could increase litter size, the effects of such genes must be thoroughly investigated if adverse effects are to be avoided. The desirability of increased litter size must be examined in the context of an animal's ability to cope with the added burden during pregnancy and lactation, and the chosen gene must be screened for detrimental effects, such as infertile females having nonfunctional ovaries.

Modification of the milk composition that might improve the quality of animal production can be brought about by transferring suitable gene constructs to cows.

The alteration of milk itself has also been discussed. The aim would be to alter cow's milk to resemble the composition of human milk more closely. Such a product might be of advantage in feeding human babies and infants.

Genetic changes which confer innate disease resistance or improve feed digestion could help eliminate undesirable effects concomitant with many current disease control practices and could profoundly alter the efficiency of food-producing animals. Recent studies suggest that a number of economically important diseases might be combated by transgenic strategies in the future. For example, transfer of the gene for a rainbow trout lysozyme which is effective against a number of economically important pathogens.

Perspective

The number of genetically engineered products at the brink of commercialisation is growing. Concerns about the consumption of recombinant food are perceivable. These concerns are less accentuated with respect to food ingredients such as enzymes but rather focus on genetically engineered organisms, such as transgenic plants and animals. In this respect, a number of questions concerning environmental and ecological risks remain unanswered. A primary concern is that transgenic plants could either become weeds which would raise the cost of weed control, or could transfer genes into wild relatives that could then develop into weeds. Other risks outlined include the inadvertent proliferation of new virus strains which could gain resistance to virus resistant plants, as well as the possible adverse impact on insects, birds, and other animals that feed on transgenic plants. [Editor's note: These concerns are discussed in the fifth article in this section.]

CAN BIOTECH PUT BREAD ON THIRD WORLD TABLES?[4]

What distinguishes Florence M. Wambugu from the 950 other researchers at Monsanto Co.'s huge life-sciences technology campus west of St. Louis are the plants crowding the climate-controlled chamber in which she works. They aren't the genetically altered strains of corn, wheat, or cotton the chemical giant hopes will become blockbuster ag biotech products. Instead, the Kenyan virologist is probing the viral susceptibility of the sweet potato. It's a minor U.S. crop, "but in countries like Kenya, it's a staple or dietary supplement—a poor man's crop where up to 50% is lost to virus," says Wambugu. "When we use biotechnology on sweet potatoes, the improvement goes to poor people."

That's the promise and the curse of ag biotech as it applies to the Third World. Initially, advocates trumpeted biotech as a tool to fight hunger. But industrial countries do most research on their own lucrative cash crops. Staples of developing nations get scant attention, because they promise scant profit. Less than $275 million has been spent on these in the past decade. And most of it has come from public or philanthropic sources.

Now, headway is finally being made. Gene mapping is leading to better rice. And Western companies, including Monsanto, Britain's Imperial Chemical Industries, and New Jersey-based DNA Plant Technology, have started programs—funded in part by international aid agencies—to train scientists such as Wambugu in ag biotech.

Few Gladiators

The need is clear. The developing world, now 70% of earth's population, will make up 90% in a generation. The World Bank

[4]Reprint of article by James E. Ellis from *Business Week* p100 D 14 '92.

says Third World food production must double in 25 years to keep pace, and Western hybrid crops and conventional farming probably can't do the job alone. Tapping biotech to increase yields or develop hardier crops could mean the difference between life or death in some poor nations by 2015.

Trouble is, most Third World nations have few biochemists and molecular biologists—the gladiators of crop engineering. Only 10 Third World nations have significant ag biotech programs. Industry won't fill the void. Most Third World countries can't afford to buy biotech products. And many don't give patent protection for Western gene discoveries or seed varieties.

That makes philanthropies and government bodies such as the U.S. Agency for International Development (AID) the purveyors of ag biotech. In the past decade, the World Bank and other public agencies have invested $180 million in the effort. And since 1985, the Rockefeller Foundation has added $50 million more. As a result, countries such as China and Thailand are making big gains in techniques needed to produce better rice strains. Biotech work on crops grown in Africa and other tropical countries is still paltry, however. "You could count the number of labs working on banana biotech on one hand," says Rockefeller Foundation Associate Director Gary Toenniessen.

There's also a lag in bringing research to market. To make that happen faster, AID has given contracts to the U.S. unit of Britain's ICI to develop insect-resistant corn for Indonesia and to DNA Plant Technology to train Costa Rican and Indonesian biotech companies in high-volume cloning of tropical crops. Both efforts include U.S. training for scientists from those countries. And the companies retain rights or royalties to any commercial products.

No Ties

Wambugu's three-year stint at Monsanto, financed by the company and by AID, aims to prepare her to introduce viral resistance into African sweet potatoes later this decade. Monsanto, which is waiving any rights to sweet potatoes that arise from Wambugu's work, also expects her to help train other Kenyan researchers. "This means we're not just tied down to [altering] sweet potatoes," says Cyrus G. Ndiritu, director of the Kenya Agricultural Research Institute, Wambugu's employer.

Transferring technology is only part of the solution. Better farming and distribution methods—often lacking in African nations—are needed to boost production of food and get it to hungry people. "Biotech will be an important tool, but it won't end world hunger," cautions John H. Dodds, managing director of the Agricultural Biotechnology for Sustainable Productivity Project at Michigan State University.

Still, for Wambugu, any progress is welcome. "Real security isn't wealth," she says, "it's food. Biotechnology can give security to poor farmers." In the end, that may be ag biotech's most important payoff.

THE COMING OF THE HIGH-TECH HARVEST[5]

Driving a pickup through the flat, fertile plains of Yolo County, California, Charlie Rominger knows every curve of the road as it meanders between an endless grid of wheat fields and the peaks of the towerlng Coast Range. He also knows the ancient history of every square inch of land. Gesturing toward what looks like a crazy quilt of grass, he explains the whole evolution of a field—Sara wheat growing on top of lentils on top of peas. Getting out of the truck, Rominger, 37, runs his hand through the dirt and pulls out a square, rusted nail. "A souvenir," he comments, "from the past."

It was back in the 1870s that Rominger's great-great-grandfather bought some acreage in the great Central Valley and launched the family farm. Since then, the rich and fertile earth has done well by the Romingers: Alfalfa and wheat, sugar beets and tomatoes, have helped to make their ranch, now 5,000 acres, one of the more successful in the state.

But the family's success is not just due to a single smart investment. Rominger, his two brothers, and their dad all hold agriculture degrees from the University of California, Davis. These sophisticated growers, as farmers in these parts are called, run their land with the help of computers and scores of publications on

[5]Reprint of article by Pamela Weintraub from *Audubon* v 92 pp92–103 Jl-Ag '92. Copyright © National Audubon Society. Reprinted with permission.

every aspect of farming, from the microbial environment to soil erosion. They also work hand in hand with scientists, providing plots where some of the latest crop varieties are tested for the very first time. The Romingers have become community leaders—in land-use issues and in new technologies. When something flies on the Rominger farm, there's a good chance it will be adopted elsewhere as well.

These days, the forward-thinking Romingers have their eyes on what some people call the most explosive advance in farming since the dawn of the agricultural age. This controversial new tool—biotechnology—is now being used in labs around the world to endow crop plants with genes from mammals, bacteria, and of course, a lot of other plants.

Biotech advocates say the new technology will increase the Romingers' wheat yield, sweeten the taste of their tomatoes, and protect their produce from disease.

Critics fear the technology could backfire: Crops transformed to resist a new generation of less toxic herbicides could discourage the family from tapping organic weed-control methods, continuing their reliance on chemicals. Crops engineered to fend off insects could become useless in a decade or less, when the bugs become immune. The Romingers themselves have a wait-and-see attitude about actually implementing the technology. But always in the market for another smart investment, they've bought some stock in the company down the road.

Calgene sits about 15 miles as the crow flies from the Rominger ranch, in the college town of Davis, just west of Sacramento. Roger Salquist, Calgene's tall and slightly on-edge CEO, pursues his vision of the future without the benefit of rolling fields or a mountain view. Instead, the company operates from a squat green-and-gray concrete building, a structure with all the prefabricated elegance of a box. In the back of this box, in a winding maze of laboratories, 65 scientists labor on crops for the 21st century—Calgene's new tomato, for instance, into which a gene was inserted that blocks the enzyme that causes tomatoes to rot. Dubbed the Flavr Savr, it resists rotting for some 10 days more than normal tomatoes. As Salquist explains, you don't have to pick them while not yet ripe to buy extra time for shipping. Instead, Flavr Savrs are left on the vine until the last possible moment, turning red and collecting all the sugars and acids that give tomatoes their rich and pungent taste. [Editor's note: The Flavr Savr tomato, the first genetically engineered food product to be made available to consumers, appeared in stores in 1994.]

The Flavr Savr is just one of Calgene's biotech products. About a mile from corporate headquarters stand nine domed farms, sealed and spectral in the Central Valley sun. These are the Calgene greenhouses. In one, reed-thin stalks of engineered canola produce seeds with oil especially low in saturated fat. In another, soft tufts of cotton withstand the onslaught of Bromoxynil, a potent herbicide that would otherwise poison the cotton as well as the weed it was meant to destroy.

Calgene is not alone. The Monsanto Company, in St. Louis, has poured hundreds of millions of dollars into agricultural biotechnology over the past decade. The company's new cotton, for instance, contains DNA from rod-shaped bacteria—called *Bacillus thuringiensis,* or B.t.—that produce protein crystals lethal to caterpillars. B.t. crystals churned out by the cotton hunker silently within until a bollworm takes a bite. Then the crystals go to work, perforating the caterpillar's stomach. The same bacteria are being engineered into potatoes and corn. Monsanto and a handful of other companies are creating crops resistant to their own herbicides—an altered harvest they claim will lower farmers' dependence on chemicals by enabling them to tap more environmentally sound and more effective herbicide brands. Scientists at institutes from UC Davis to the University of Ghent, in Belgium, meanwhile, are creating crop plants resistant to drought, salt, and disease. [Editor's note: The many different ways in which recombinant DNA technology has been applied to the production of food are detailed in the following article.] And the Rockefeller Foundation, in New York City, is investing millions of dollars in a lofty project of its own: rice engineered to resist disease and provide plentiful nutrients for the exploding populations of the developing world.

Alvin Young, director of the Office of Agricultural Biotechnology at the U.S. Department of Agriculture (USDA), believes the technology for the transformed crops comes in the nick of time. "World population will probably double in the next forty years," declares Young, "forcing us to produce twice as much food on the same amount of land." The solution? "A global gene pipeline," according to Young, that delivers the seeds of plants engineered to thrive in precise locales. He goes on, "We will have cassava plants tailored for India and cassava plants tailored for Kenya. We will engineer plants that can thrive under tremendous regional pressures, from drought to chemical pollution to the onslaught of cold. By the year 2010 the technology will be pervasive, because it is based on the ability to manipulate biologically based systems at the ultimate level—the gene." [Editor's note: Some of the challenges associated with trying to alleviate hunger in the Third World by developing genetically engineered food crops are discussed in the fifth article in this section.]

But many are disturbed by the awesome power of the technology. Michael Picker, head of Sacramento's National Toxics Campaign, says that when molecular biologists alter genes, they may be changing organisms in ways that will not be truly known for years. "Just one handful of soil contains billions of interacting bacteria," Picker explains. "When we dramatically shift the genetic makeup—and the functioning—of a single organism, how do we know it won't affect the whole chain?"

Critics also fear the new technology will tie us ever more tightly to what they call silver-bullet solutions—one-shot cures based on chemicals and genetically engineered organisms that must be produced and supplied by industry on a continual basis to keep a farm going. According to Jane Rissler, a plant pathologist and biotechnology specialist at the National Wildlife Federation, such solutions place control over agriculture—and food production as a whole—in the hands of companies interested only in expanding market position, not in helping humankind. "The more research money we pour into these silver-bullet solutions," she says, "the less likely we'll be to find other, more sustainable means of controlling crop disease and weeds."

Rissler believes it's no accident that many of the researchers in this field—she calls them gene-jockeys—are men. "They get a thrill out of creating life," she says. "I know a man out in California who talks about building potatoes. He's going to build potatoes by adding genes. What arrogance! Man, you've already got a potato. You're just tinkering!"

Of course, both men and women have tinkered with crops since the beginning of agriculture. When our ancestors left the forest for the open plains some 40,000 years ago, they survived as hunter-gatherers, picking fruits and berries and trekking after game. When they finally domesticated plants and animals in their own backyard, they learned to nurture those that were hardier or more fertile so they could produce a little more.

In the 1940s, the powerful arm of science revolutionized agriculture to feed the ever-expanding population of the world.

While war raged across the globe, American agronomist Norman Borlaug, father of the first green revolution, worked in the fields of Mexico. Crossbreeding wheat, he developed high-yield crops far more resistant than standard varieties to disease and weather damage. Borlaug's extraordinary work helped to increase food supplies in Mexico and throughout Asia.

But while Borlaug and colleagues managed to increase food production, in many cases the environment paid a heavy price. The bold new crops were able to grow only with the help of chemical fertilizers and pesticides, along with controlled irrigation and drainage. In 1962 biologist Rachel Carson exposed the devastating impact of such chemicals in her classic, *Silent Spring*. Carson pointed out that farm chemicals such as the pesticide DDT were draining into streams and rivers, killing fish, plants, and the animals that fed on them. Soon the science of engineering the new agricultural order—the one based on all those chemicals—was at war with the science of protecting the environment.

In 1973 an enormous scientific advance seemed to herald a truce between the two camps. A couple of California scientists, geneticist Stanley N. Cohen of the Stanford University School of Medicine and biochemist Herbert W. Boyer of the University of California, San Francisco, developed technology for transferring foreign genes into bacteria. In a splashy display of the technique, the team used molecular "scissors" known as restriction enzymes to snip genes out of the chromosome of a toad cell. Then they inserted a toad gene into a plasmid, a small packet of DNA able to sneak genetic information into foreign bacteria. Soon a whole population of bacteria had begun to incorporate and reproduce toad genes, becoming some of the first critters ever based on the breakthrough technology called recombinant DNA.

If plants could be transformed like this, researchers started saying, then it would not be long before we could engineer crops with genes for almost any characteristic at all. Who would need pesticides when plants could incorporate genes conferring resistance? And who would need chemical fertilizer when crops with internal nitrogen fixation genes could create their own? At last they had a beneficent science that might blast chemicals from the agricultural scene.

But Monsanto, one of the first multinational companies to invest heavily in agricultural biotech, had another idea. Aware that the technology could be used to endow plants with new genes—and new traits—the company saw a means of bolstering its own revenues, which at the time were threatened by attacks on farm chemicals.

Under the direction of an energetic biologist, the late Howard Schneiderman, Monsanto built the $165 million Life Sciences Research Center on 210 acres west of St. Louis. By 1983 two of the hundreds of researchers hired by Schneiderman—Robert T. Fraley and Robert Horsch—had created the world's first "transformed plant," a petunia that incorporated the genes from a bacterium. The race to create engineered products for tomorrow's farm had begun.

Rebecca Goldburg, now chair of the biotechnology program at the Environmental Defense Fund (EDF), in New York City, began to scrutinize such products in 1986. Jeremy Rifkin, president of the Washington-based Foundation on Economic Trends, voiced strong opposition to the release of genetically engineered microbes. The most publicized of these, the notorious Ice-Minus, was a strain of bacterium genetically altered so it would no longer produce the protein that causes dew to freeze when temperatures hit between 25 and 30 degrees Fahrenheit. The idea was to coat strawberries, potatoes, and other crops with Ice-Minus, crowding out naturally occurring bacteria and giving the plants an extra measure of frost protection.

Steven E. Lindow, a plant pathologist from the University of California, Berkeley, who directed the project, insisted the release of Ice-Minus was inherently safe. In a background report issued at the time, the university itself said that "neither the commonly occurring bacteria, nor the modified ones, are harmful to humans or animals. The modified bacteria are nearly identical to the strains found on crops and other plants everywhere. The only difference is that they lack the single gene that allows ice to form on plant leaves. Such variations," the university added, "occur in nature, so the strain being tested is 'new' only in [terms of] the technique used to make the change. No new traits have been added."

Rifkin disagreed, pointing out that virtually no research had been done on the long-term effects of genetic engineering. "People will pay for this hundreds of thousands of years from now," Rifkin said at the time. "Every introduction is a hit-or-miss ecological roulette."

As for Goldburg, she was more concerned about future uses of the technology than about Ice-Minus itself. And she felt that both camps were naive. "The genetic engineers claimed that nothing they did could in any way be risky," she recalls, "in part because everything that could have happened to microorganisms had already been tried by evolution itself. But even if a mutation has occurred before, that's not the same as putting millions of altered microbes in a new environment in which those organisms will thrive."

Unlike Rifkin, however, Goldburg was not concerned about creating monsters but about creating pests. "I didn't think we'd see the construction of an Andromeda strain," she says, "but just some new organism that would be costly, something that might make it hard to maintain natural areas in a pristine state."

As the 1980s rolled around, one new technology bothered Goldburg the most. In field tests around the nation and the world, chemical companies were starting to pioneer cotton and soybeans, tomatoes and tobacco, engineered to resist the companies' own herbicides—a new generation of weed killers the corporations claimed were far less toxic than herbicides used before.

As Goldburg saw it, since the resistant new crops could grow in the presence of amounts of herbicide that would harm or kill nontolerant crops, there would be little incentive for farmers to control their herbicide use. The growers of the world would become tied ever more tightly to the cycle of chemicals, missing out on the promise of sustainable agriculture made so many years before.

Goldburg and colleagues also worried that the resistant new crops would pollinate closely related weedy species, thus passing on herbicide resistance to their weedy relatives. Whether or not this will actually occur is still a matter of debate. Chemical companies cite studies showing that if crops are controlled and kept far from weedy relatives, genetic drift will be insignificant. "We've conducted a large-scale study to see if genes for herbicide resistance flowed out of our cotton plants into other fields," says Calgene's Salquist, "and found that if there's a certain amount of space between fields, they do not."

Yet a review of professional journals shows that many scientists are not convinced. Ecologist and evolutionary biologist Kathleen H. Keeler of the University of Nebraska points out that a weedy race of millet seems to have evolved just recently in Wisconsin and Minnesota—after 200 to 300 years of millet cultivation in North America without weed problems. "Until such events can be anticipated," Keeler says, "there will be an ongoing risk of weeds derived from genetically engineered crops."

With so much still unknown, the scope of the new herbicide-resistant crops has nonetheless grown vast. Working with France's Rhône-Poulenc Agricultural Company, Calgene has developed cotton resistant to the herbicide Bromoxynil. German's Hoechst is engineering maize resistant to its herbicide Basta. And the USDA itself is developing plants resistant to 2,4-D, a close relative of the defoliant Agent Orange and a common ingredient in many agricultural applications. These groups and others are engineering herbicide resistance into virtually all major food crops, from rice and corn to potatoes and wheat.

Some of these herbicides may pose dangers to farm workers and the population at large. According to the Environmental Protection Agency (EPA), for instance, a recent study shows that pregnant rats exposed to Bromoxynil either orally or through skin contact bear offspring with defects. The agency is so concerned about this herbicide that it now requires all workers who load, mix, or apply the chemical to wear protective garb. A 1990 study by the National Cancer Institute shows that the common weed killer 2,4-D tripled the risk of cancer of the lymph nodes in a group of Nebraska farmers. The use of Atrazine, already criticized for polluting groundwater in California's Central Valley and in Los Angeles, has been restricted in parts of the state.

Chemical companies are also extending their markets in pesticides. A case in point: Monsanto's B.t. cotton, which incorporates bacterial gene coding for protein crystals lethal to bollworms. The product should be ready for market, says Monsanto's Fraley, by the mid-'90s. Scientists at Monsanto and elsewhere are also engineering B.t. genes into potatoes, corn, and other crops.

Ed Bruggemann, a molecular biologist with the National Audubon Society, says, "Engineered plants have the ability to reduce the use of chemical insecticides. Indeed, when B.t. found in nature is simply sprayed on plants as a natural pesticide, resistance results. With B.t. engineered into the plant, exposure will be greater and the force to evolve resistance more intense. Insects can develop resistance to B.t. just as they develop resistance to chemical insecticides. The concern is that this technique will work for just a few years, then farmers will have to return to chemical pesticides." Bruggemann recommends that companies look for ways to prevent the new strains from quickly becoming obsolete.

"Trouble is," he says, "companies have little incentive to extend the life of the product because they can hold the patent for only seventeen years. It may be to the corporation's great advantage to have old products die so new ones can come to the fore. Corporations would rather sell more of the product at the start, and get their money back as soon as they can. B.t. crops will last three to four years if we use the product poorly, and thirty to forty years if we use the product well."

Fraley says Monsanto shares Bruggemann's concerns and has a resistance-management program firmly in place. "We're investigating a whole spectrum of strategies," he states, "to prolong the usefulness of our B.t. crops." For instance, the company is combining B.t. pesticides with limited application of traditional chemical pesticides. It's also trying a technique known as integrated pest management, in which planting and harvesting are timed to exploit natural predators to help eliminate pests. Finally, company scientists plan to develop different B.t. varieties so that if pests become immune to one, another will be ready to take its place.

The final product in the process—genetically engineered food—comes with some promise. There is Calgene's engineered canola plant. Monsanto is gearing up to produce potatoes with greater starch content, which absorb less oil and fat, to produce healthier chips and fries. And Louisiana State University scientists are developing nutritious forms of rice with storage proteins from beans and peas.

But Goldburg feels it's impossible to know the impact of gene changes in food without intensive analysis. Engineered foods could be a concern for people with allergies and could play havoc with religious dietary laws. After all, even a passing glance at the field test applications on file at the USDA reveals potatoes with chicken and insect genes, walnuts with bacterium genes, and rice with genes from corn.

Roger Salquist says that Calgene, for one, has analyzed the Flavr Savr for toxins and changes in nutrient content and found none. But Goldburg warns that other companies may follow through only if adequate regulations are in place. In a recent report entitled *A Mutable Feast: Assuring Food Safety in the Era of Genetic Engineering,* Goldburg and other EDF staffers call for a new roster of rules: If genetically engineered food contains a new substance, the Food and Drug Administration (FDA) should regulate and label it like any other product with an additive. And all such foods should be analyzed for elevated levels of naturally occurring toxins or decreasing levels of nutrients, just to make sure they're appropriate and safe.

As multinational corporations like Monsanto and Ciba-Geigy begin to integrate development of the seeds, the agrichemicals, and possibly the food itself, they may achieve a new level of control over the agricultural resources of the world. Already, says Jack Doyle, Director of the Agricultural and Biotechnology Project at Friends of the Earth, these huge corporations have begun to buy up smaller genetic engineering and seed companies. The result, he believes, will be a "life sciences conglomerate," an unprecedented institution of enormous economic and political power.

These megacompanies, says Doyle, "are using genes just as earlier corporate powers used land, minerals, or oil. In many ways, DNA is the ideal corporate resource: It can be patented and wielded as property. It can be manipulated in the laboratory. It can replace or reduce reliance on cumbersome raw materials like farmland or feedstocks, reduce labor costs, and circumvent finicky variables such as weather. Finally, DNA can be used to produce tremendous quantities of rare and expensive products for pennies." As Doyle points out, the mergers and buyouts in the biotechnology arena do not represent a new form of efficiency or economic vibrancy with the potential to help humankind. Rather, the technology is being wielded so that companies may extend current market positions and establish others.

This trend is exacerbated, says UC Davis rural scientist Martin Kenney, because even university scientists are receiving more funding from industry. According to Kenney, author of *Biotechnology: The University-Industrial Complex,* "Large companies like Monsanto fund university research programs at up to four hundred thousand dollars per shot. All you have to do is read the newspapers to see that in other cases professors are getting massive blocks of stock from companies they consult for." In effect, Kenney adds, industry has directed its funds so that university scientists do the basic molecular biology while the company itself develops the seed. "Thus, the university is not just providing seed free to small seed firms and farmers, as was done in the past. Instead, large companies create the seed and link it, at the genetic level, with a chemical. The companies set the agenda and become the central conduit in the production of our food."

With clear and balanced regulations, these problems and others might be kept under control. But the regulations governing biotechnology, say many experts, are tangled and obscure. To gain approval for a field test, researchers must apply to the USDA. If their product is considered a pesticide (B.t. cotton, for instance) or a toxin released into the environment (Ice-Minus), they must apply to the EPA as well. Once the product is ready to be marketed as a food (Calgene's tomato), it also falls under the domain of the FDA.

David MacKenzie, director of the USDA's National Biological Impact Assessment Program, says that many independent researchers are discouraged from conducting field tests because wading through the regulations is such a chore. "Companies like Monsanto," says MacKenzie, "have employees who work full-time just negotiating the regulatory maze. As a result, many profit-making applications move forward while more beneficial projects never see the light of day."

While the regulatory maze is nothing new, biotechnology is. And, say the critics, this striking new technology should be governed by laws of its own. But a new Bush administration policy, written by the staff of Vice-President Quayle and his Council on Competitiveness, holds that genetically engineered products are not intrinsically dangerous and that they deserve no more scrutiny than products created in a more conventional way. Regulatory review of biotech, the administration now says, should be "designed to bring products to market without too many roadblocks put in the way."

"This four-billion-dollar industry should grow to fifty billion by the end of the decade—if we let it," President Bush recently told the U.S. Chamber of Commerce. "The United States leads the world in biotechnology, and I intend to keep it that way." [Editor's note: The Bill Clinton administration's approach to plant biotechnology is not substantially different from that pursued by the Bush and Reagan administration.]

To that end, the FDA announced in May that it will approve genetically engineered foods without considering them inherently dangerous or requiring extraordinary levels of testing—unless special safety issues, including the problem of food allergies, arise. According to the White House, the FDA policy will serve as a model for officials at the USDA and the EPA, two other agencies involved in regulating biotech.

The Bush administration's approach disturbs Goldburg, who finds that it leaves a policy "so vague that the Office of Management and Budget will be able to block any regulation it wants. I foresee a regulatory vacuum as a result."

But Terry Medley, director of Biotechnology, Biologics, and Environmental Protection at the USDA, says the current regulations are completely sufficient to regulate biotech products: "Our role is to make the most rational, the most informed, decision by scientifically assessing the risks using all the information we have at hand. And that is something, given our expertise, we can do."

In fact, Medley and other USDA officials seem to be on a biotech mission. "You've heard of Star Trekkers," Medley says. "Biotekkers are people involved in biotechnology." It's no surprise that Medley brings with him not just the regulator's narrow focus, but also the visionary's zeal. "We believe this technology will pave the way to sustainable agriculture," he says, "reducing our reliance on chemicals and providing farmers with choices that can cut their costs. Over the long term, we'll create value-added crops—crops with higher nutritional value, crops that grow despite cold or drought. This is a global issue. The regulations we set here will help to establish standards worldwide." Medley says those international standards will come in especially handy for nations with the least sophistication, those of the Third World.

Gary Toenniessen, the Rockefeller Foundation's associate director of agricultural sciences, says that "in the next century, Third World countries will need to grow increasingly more food on the same amount of land. Yet the existing technology has already pushed rice production as high as it can. American and European corporations had little interest in this effort, which didn't have much profit potential in the developed world." Rockefeller-supported researchers at institutions as diverse as the University of Nottingham, in the United Kingdom, and Scripps Research Institute, in San Diego, are engineering rice that resists viral disease, withstands drought, and produces a higher yield. According to Toenniessen, a major thrust of the effort involves training Third World scientists from nations like Thailand, Nepal, and Bangladesh to modify and implement the technology themselves.

Robert Herdt, director of the foundation's agricultural program, meanwhile, is overseeing the environmental and social impacts of the technology. "We know the technology can be used in the wrong way, and it can make environmental problems worse," Herdt says. "On the other hand, we can direct our program so that these new crops do away with irrigation systems or pesticides that disrupt the environment. We won't support herbicide-resistant crops, because pulling weeds is a major source of income in the Third World. If herbicides killed the weeds, there would be far fewer jobs and we would be shifting money into the chemical companies and away from the poor."

Whether the focus is the developed or the developing world, of course, there's no guarantee that the safeguards Herdt envisions will be enforced. Bob Cantisano, a consultant on organic and sustainable farming, echoes the views of many when he says that if unchecked, "biotechnology will displace the farm community. By selling the seeds and the chemicals those particular seeds require, major corporations will concentrate agricultural wealth. Industry has no incentive to promote an agriculture with less chemical input, yet the input is now becoming so expensive that farmers can't survive. Biotechnology will mean more input and will further stress the small family farm."

Pointing to a vineyard unattended, a wheat field gone fallow, Charlie Rominger agrees that more farmers go under each and every year. "Those who farm like their grandfathers farmed," he says, "are winnowed out each cycle." Striving to stay economically and environmentally sound, the Romingers are growing organic tomatoes and are using USDA-imported wasps from Tashkent to control the aphids nibbling their wheat. Yet, Charlie Rominger concedes, biotechnology will probably become part of the arsenal he wields in sustaining his inheritance, the family farm. "We'll look at this technology carefully," he says. "We'll examine what others do first. But it looks like biotechnology will help us stay competitive, and we've got to stay competitive if we want to survive."

NO HUMAN RISKS: NEW ANIMAL DRUG INCREASES MILK PRODUCTION[6]

Milk. It's been a staple for ages among children, adolescents and adults. We're told milk helps build strong teeth and bones and healthy bodies.

Over the last several years, scientists have been developing genetically engineered products to help dairy cows produce more milk.

[6]Reprint of article by Kevin L. Ropp from *FDA Consumer* v 28 pp24–7 My '94. Reprinted with permission.

Last Nov. 5, the Food and Drug Administration approved one such product, sometribove (Posilac), Monsanto Co.'s genetically engineered bovine somatotropin (bST).

Genetically engineered, or recombinant bST is virtually identical to a cow's natural somatotropin, a hormone produced in its pituitary gland that stimulates milk production. The primary difference between the two is that rbST may include additional amino acids. Injecting rbST can increase a cow's milk production by 10 to 15 percent.

"This has been one of the most extensively studied animal drug products to be reviewed by the agency," said FDA Commissioner David A. Kessler, M.D. "We examined more than 120 studies. There were several advisory committees. The public can be confident that milk and meat from bST-treated cows are safe to consume."

At press time, FDA was still reviewing rbST products from American Cyanamid, Eli Lilly & Company, and Upjohn Company. "Each of the remaining products will be reviewed as stringently as the first," said Suzanne Sechen, Ph.D., an animal scientist in FDA's Center for Veterinary Medicine's division of biometrics and production drugs.

Recombinant bST is made in much the same way as synthetic human insulin for treating diabetes. The gene for bST is inserted into special bacteria, which then reproduce, replicating the gene. During manufacture, the bST is collected from these bacteria and further processed. The finished, sterile product is then formulated for use in dairy cows.

FDA Review

Recombinant bST first came under FDA review in the early 1980s, when the four companies submitted investigational new animal drug applications. Since that time, the agency has authorized rbST testing on more than 20,000 cows in the United States.

During FDA's review of rbST, there has been much public debate on safety and economic issues related to the drug. Some organizations opposed to the use of rbST have contended that it causes health problems for cows injected with the drug, for their calves, and for humans who consume milk or meat from these animals.

"We went to unprecedented lengths to resolve every issue raised," said Richard Teske, D.V.M., acting director of FDA's Center for Veterinary Medicine, "not only through our own rigorous review process, but also by subjecting our findings to peer review through a published journal article and by an outside committee of experts."

The agency's conclusion—that rbST poses no risk to human health—has been affirmed by scientific reviews in the last several years by the National Institutes of Health; the Congressional Office of Technology Assessment; drug regulatory agencies of Canada, the United Kingdom, and the European Union; and by the Department of Health and Human Services' Office of the Inspector General.

FDA's Findings

Under the Federal Food, Drug, and Cosmetic Act, FDA is responsible for ensuring the safety to humans of the milk, meat, and other food products from food animals treated with a new drug, as well as the safety and effectiveness of the drug for the animals. The agency also must ensure that the manufacture and use of the product does not pose environmental hazards.

FDA's Center for Veterinary Medicine determined in the mid-1980s that food products from rbST-treated cows are safe for human consumption. In the Aug. 24, 1990, issue of *Science,* FDA scientists summarized more than 120 studies that examined the safety of milk and meat from dairy cows treated with rbST, concluding that use of rbST presents "no increased health risk to consumers."

The agency's determination was based on a number of scientific findings. Because it is a protein-based hormone, rbST is broken down during digestion, which renders it biologically inactive and incapable of having any effect in humans or animals. Even if injected in humans, bST has no effect, according to studies done in the 1950s that looked at natural bST as a possible treatment for human dwarfism. (It didn't work.)

In addition, studies show that pasteurization destroys approximately 90 percent of bST, natural or otherwise, present in milk.

Also, studies have shown that rbST does not affect the nutritional qualities of milk. Scientists are unable to detect a difference between milk from rbST-treated cows and from untreated cows.

Summing up the public health issues surrounding rbST, Kessler said, "there is virtually no difference in milk from treated and untreated cows. In fact, it's not possible using current scientific techniques to tell them apart."

NIH Review

In 1990, a special panel of the National Institutes of Health also looked at recombinant bST. Its members unanimously concluded that rbST is effective in increasing milk production and that the composition and nutritional value of the milk from the treated cows are essentially the same as from untreated cows.

In addition, the panel found that well-managed, rbST-treated cows experience no greater health problems than untreated cows of equal production and that calves from cows administered rbST have normal birth weights, growth, and development.

FDA also reviewed scientific data to see how rbST affects cows. It concluded that rbST causes no serious or long-term health effects in treated cows or their offspring.

Mastitis and Antibiotics

However, FDA found evidence in the clinical trial data submitted by Monsanto that cows treated with sometribove have a slightly increased incidence of mastitis, a common infection of the udder. A September 1992 report by the General Accounting Office raised concerns that, because mastitis is often treated with antibiotics, sometribove use could lead to increased antibiotic residues in milk or meat. This could pose a potential health problem, for example, to people allergic to the antibiotics who consume the meat or milk.

In March 1993, FDA brought these concerns to an advisory committee. The committee concluded that the effect of sometribove use on the incidence of mastitis was much less than other factors, such as the season, age of the cows, and herd-to-herd variation. They also concluded that adequate safeguards are in place to prevent unsafe levels of antibiotic residues from entering the milk supply.

Drugs such as antibiotics may be used in food-producing animals only under government-approved conditions and with appropriate withdrawal periods, as established by FDA, to ensure that the food is safe for people to eat.

Ensuring Milk Safety

Milk safety is ensured through a joint effort of FDA and the National Conference on Interstate Milk Shipments, an organization of state health officials and members of the dairy industry. The conference oversees a Voluntary Cooperative State-Public Health Service Program for Certification of Interstate Milk Shippers (IMS) to administer milk safety rules. Responsibilities under this program are divided between state agencies with FDA oversight.

The states are the primary operators of the IMS program, with FDA providing scientific, technical and inspection assistance, as well as auditing state programs. All 50 states, the District of Columbia, and Puerto Rico participate in IMS, and nearly 150,000 dairy farms and 800 milk plants, representing almost 90 percent of the total milk production in the United States, are covered.

Milk tanker-trucks may stop at five to 10 different farms to collect a load of raw milk. A sample of each farm's milk is collected, labeled and stored in a special section on the truck. The rest of the milk is loaded into the truck's usually common milk tank.

When the tanker is full, the load is delivered to a milk processor. The dairy industry currently tests all tankers for penicillin-like beta-lactam drugs before processing the milk. Beta-lactams are the most commonly used drugs for treating mastitis.

If a load is found to have unsafe residues, all the milk must be discarded. The individual samples collected from each farm are then tested to determine the source of the residue. Producers responsible for unsafe residues are subject to regulatory sanctions that may include suspension or revocation of permits or monetary fines.

In addition, once every three months state inspectors randomly sample 10 percent of the tanker-trucks coming into milk processing plants. The states also randomly collect samples from every farmers' milk storage tank at least four times every six months.

Because of these safeguards, FDA's Veterinary Medicine Advisory Committee concluded that the increased risk to human health posed by mastitis and the use of antibiotics is insignificant and manageable.

Additional Monitoring

Nevertheless, at FDA's request, Monsanto has agreed to take additional steps to ensure that sometribove use does not lead to an increase in antibiotic residue levels.

"In addition to all of the studies, and the current nationwide milk monitoring system, we have put in place an extensive post-approval marketing program that will assure that food products from bST-treated cows meet the high standard of safety required by our statute and demanded by the public," Kessler said.

Monsanto will conduct a post-approval monitoring program that includes:

- a two-year tracking system of milk production and drug residues in top dairy states, representing over half of the nation's milk supply, that will periodically compare the amount of milk discarded after sometribove is marketed to the amount discarded before approval
- a 12-month comparison of the proportion of milk discarded due to positive drug tests between sometribove-treated and untreated herds
- a reporting system to monitor all sometribove use and follow up on all complaints
- a sampling of 24 commercial dairy herds using sometribove with specific monitoring for mastitis, animal drug use, and the resulting loss of milk.

Labeling

Labeling milk and other foods from rbST-treated cows was another issue FDA considered during its review. Last May, the agency's food and veterinary medicine advisory committees met for two days to discuss labeling. Based on the committee members' conclusions and on its own review, FDA concluded that it lacks a basis under the statute to require special labeling of foods from rbST-treated cows. Food companies may voluntarily label their products as long as the labeling is truthful and not misleading.

Recently, several states, industry, and consumer representatives have asked FDA for guidance on voluntary labeling of milk —both from cows that have been treated with rbST and those that have not. The agency issued interim guidance Feb. 8, which should help ensure that consumers receive truthful and non-misleading information about milk and milk products.

The guidance states: "Because of the presence of natural bST in milk, no milk is 'bST-free,' and a 'bST-free' labeling statement would be false.

"Also, FDA is concerned that the term 'rbST-free' may imply a compositional difference between milk from treated and untreated cows rather than a difference in the way the milk is produced. Instead, the concept would better be formulated as 'from cows not treated with rbST' or in other similar ways. However, even such a statement, which asserts that rbST has not been used in the production of the subject milk, has the potential to be misunderstood by consumers. Without proper context, such statements could be misleading. Such unqualified statements may imply that milk from untreated cows is safer or of higher quality than milk from treated cows. Such an implication would be false and misleading."

In response to consumers' safety concerns, FDA's Teske sums it up: "FDA's review of a new animal drug is a very rigorous process. When a product has completed that process and our conclusion is that it's safe and effective, we have a great deal of confidence in our decision. We feel the public can have that same degree of confidence as well."

BIBLIOGRAPHY

An asterisk (*) preceding a reference indicates that the material or part of it has been reprinted in this compilation.

Books and Pamphlets

Annas, George J. & Elias, Sherman, eds. Gene mapping: using law and ethics as guides. Oxford University Press. '92.

Barth, Joan C. It runs in my family: overcoming the legacy of family illness. Brunner/Mazel. '93.

Bishop, Jerry E. & Waldholz, Michael. Genome: the story of the most astonishing scientific adventure of our time—the attempt to map all the genes in the human body. Simon & Schuster. '90.

Cook-Deegan, Robert M. The gene wars: science, politics, and the human genome. Norton. '94.

Cranor, Carl F., ed. Are genes us? the social consequences of the new genetics. Rutgers University Press. '94.

Crick, Francis. What mad pursuit: a personal view of scientific discovery. Basic. '88.

Doyle, Jack. Altered harvest: agriculture, genetics, and the fate of the world's food supply. Penguin. '86.

Fletcher, Joseph Francis. The ethics of genetic control—ending reproductive roulette: artificial insemination, surrogate pregnancy, nonsexual reproduction, genetic control and screening. Prometheus. '88.

Frankel, Mark S. & Teich, Albert H., eds. The genetic frontier: ethics, law, and policy. American Association for the Advancement of Science. Directorate for Science & Policy Programs. '94.

Kevles, Daniel J. & Hood, Leroy E., eds. The code of codes: scientific and social issues in the Human Genome Project. Harvard University Press. '92.

Lee, Thomas F. The Human Genome Project: cracking the genetic code of life. Plenum. '91.

Levine, Joseph S. & Suzuki, David T. The secret of life: redesigning the living world. WGBH Boston. '93.

Nelson, J. Robert. On the new frontiers of genetics and religion. Eerdmans. '94.

Shapiro, Robert. The human blueprint: the race to unlock the secrets of our genetic script. St. Martin. '91.

Silverstein, Alvin & Silverstein, Virginia B. Genes, medicine, and you. Enslow Publications. '89.

Thompson, Larry. Correcting the code: inventing the genetic cure for the human body. Simon & Schuster. '94.

Watson, James D. The Double Helix. '68.

Weir, Robert F.; Lawrence, Susan C.; & Fales, Evan, eds. Genes and human self-knowledge: historical and philosophical reflections on modern genetics. University of Iowa Press. '94.

Wilkie, Thomas. Perilous knowledge: the Human Genome Project and its implications. University of California Press. '93.

Wingerson, Lois. Mapping our genes: the Genome Project and the future medicine. Dutton. '90.

Additional Periodical Articles with Abstracts

For those who wish to read more widely about the impact that the advances in genetic research have had on modern society, this section contains abstracts of additional articles that bear on the topic. Readers who require a comprehensive list of materials are advised to consult the *Readers' Guide to Periodical Literature* and other Wilson indexes.

Genetics meets forensics. Ricki Lewis. *BioScience* 39:6–9 Ja '89

The technique of DNA printing, or DNA fingerprinting, is increasingly being used as evidence in court trials. The method, which can establish a person's identity by detecting inherited DNA sequences, has been used to compare crime scene material and tissue from the victim and the accused, to determine paternity, and to settle immigration disputes by assessing relatedness between purported relatives living in two nations. The FBI is assessing three types of DNA printing—single- and multi-locus probes and the dot-blot method—to determine which combination best suits particular forensic situations. These approaches differ in the clarity and sensitivity of the end result and in the amount of tissue required. The use of such evidence in court has caused a knowledge gap; many judges and juries are unfamiliar with the biology on which the method is based. Conversely, many biologists do not know the legal hurdles that new evidence must pass before being accepted by a court of law.

DNA-based identity testing in forensic science. Kevin C. McElfresh, Debbie Vining-Forde, and Ivan Balazs. *BioScience* 43:149–57 Mr '93

For more than 5 years, DNA-based identification testing in forensics has been used in courts, but the technology and its application have been repeatedly challenged. An examination of the fundamentals of DNA-based identification testing, the current state of the technology, and its use in court reveals that the method is scientifically sound. In terms of providing the maximal amount of probative information on which innocence or guilt can be evaluated by a jury, DNA-based forensic analysis of crime scene samples proves significantly more informative than any previous biological technique. The ethical aspects of DNA fingerprinting, various procedures used in DNA forensics technology, criteria used to determine a match between DNA fragments, the phenomenon of band shift and its possible influence on forensics results, the role of population genetics in DNA fingerprinting, and the future of DNA-based forensics are discussed.

A historical view of social responsibility in genetics. Jonathan Beckwith. *BioScience* 43:327–33 My '93

Geneticists should heed the lessons of history and endeavor to prevent their discoveries from being used to justify harmful social policies. In the early 20th century, the leaders of the eugenics movement cited the new concepts of genetic science to support their claims that certain races, ethnic groups, and social classes were inferior to others. Many prominent geneticists supported the eugenics movement during its early years. Even after it became clear that the so-called science of eugenics lacked legitimacy, few geneticists spoke out against it. The idea that genes govern human behavior fell into disfavor after World War II, but it began attracting attention again in the late 1960s. Recent progress in DNA sequencing has prompted some biologists to suggest that genetics can explain almost all biological processes and social problems. This focus tends to distract society's attention from other possible explanations and solutions.

A storm is breaking down on the farm. Joan O'C Hamilton. *Business Week* 98–101 D 14 '92

Calgene Inc. of Davis, California, intends to introduce tomatoes that have been genetically engineered to retard rotting, but a consumer backlash could be in the offing. Because they will not rot during shipping, these tomatoes can be picked riper and should be redder and better tasting than other tomatoes, which are picked green and artificially ripened. Eventually, the new tomato may clear the path for several other foods genetically engineered to resist rotting, pests, and disease. The tomato, however, has come under attack by some consumer and environmental activists, as well as some growers and restaurateurs, who feel that its biotech origins make it inherently unsafe. The article discusses the work of well-known agricultural biotech foe Jeremy Rifkin, who has vowed to quash the new tomato; the method through which the new tomato was created; and the potential windfall for investors should bioengineered food products succeed.

Mean green. Joel Keehn. *Buzzworm* 4:32–7 Ja/F '92

Despite the media hype surrounding genetic engineering's ability to transform agriculture, people involved in sustainable agriculture are concerned about the current direction of biotechnology. Supporters of the genetic engineering movement tout the technology as a solution to world hunger. This vision, however, remains an illusion, because no biotechnology company is likely to produce crops that can survive drought or fix nitrogen. Instead, these companies are developing herbicide resistant crops that can withstand higher doses of the chemicals that the same companies produce. Proponents of sustainable agriculture worry that such an approach will only perpetuate American agriculture's reliance on chemicals and continue to place control of agriculture in the hands of large corporations. They are also concerned that genetically engineered plants could cause even more problems in developing countries by cross pollinating with wild relatives.

Gene research considered. *The Christian Century* 109:512 My 13 '92

"Genetics, Religion and Ethics," a recent meeting sponsored by the Institute of Religion at the Texas Medical Center and by Baylor College of Medicine, brought together some 200 scientists and theologians to discuss the ethical and religious issues raised by genetic manipulation. Citing the complexity of the issues, the participants decided to move slowly in issuing a statement on how religion might shape future health and pastoral care in the "brave new world" represented by genetic advances.

Roundtable: the Human Genome Project. *Issues in Science and Technology* 10:43–8 Fall '93

The successes of the Human Genome Project, an international effort to locate and catalogue every gene, are creating new ethical and social policy dilemmas. Genome researchers are unlocking the mechanism of inherited disease, but the ability to identify people who have or carry such illnesses inevitably precedes the discovery of treatment. In a forum held at the University of Texas Southwestern Medical Center on May 6, experts addressed such topics as the role of Congress in regulating the use of individual genetic data, how to educate legislators and the public about genetics, the possibility of regulatory guidelines, insurance reform, screening for genetic abnormalities before conception, how to present a balanced view in genetic counseling, the National Institutes of Health's recent decision to halt funding for clinical genetics training programs, international cooperation in genetics research, and clinical observation of families.

Nearing the final frontier. Tom Fennell. *Maclean's* 104:37 Jl 15 '91

Part of a special section on gene research. Genetic engineering has progressed to the point that researchers can change the genetic makeup of a human embryo. This so-called germ-line therapy has raised many ethical questions because it interferes directly in the development of offspring and because any changes made would recur in future generations. Critics contend that the boundary between gene-implant therapy, which would treat genetic diseases without passing genetic changes to future generations, and germ-line therapy could easily be crossed. In 1987, a Medical Research Council committee chaired by Patricia Baird, head of the department of medical genetics at the University of British Columbia, issued a series of guidelines aimed at monitoring genetic research in Canada. The guidelines, which have not been backed up by legislation, sanction gene-implant therapy but oppose germ-line experiments.

Family ties: the health and heredity link: knowing your relatives' medical past can help protect your physical future. Jill Neimark. *Mademoiselle* 99:196+ My '93

Because many diseases and health conditions are hereditary, it is crucial to know your relatives' medical histories. Such knowledge can promote early diagnosis of possible problems and can even help people prevent disease by changing eating, exercise, and other personal habits. Questions are suggested for inquiring into relatives' medical backgrounds.

Inherent risks. Richard Laliberte. *Men's Health* 7:58–60+ N/D '92

Research is proving that many health problems can be hereditary. Scientists have found that humans carry as many as 20 potentially harmful genes that could trigger or pass along inherited health problems, and they have begun the long process of determining precisely which genes are responsible for which problems. By mapping all the genes in the human body, researchers hope to be able to develop accurate screening tests to point out genetic glitches and new drugs that will halt or block the actions of flawed genes. The results of gene mapping may not be evident for decades, however. In the meantime, researchers recommend that people safeguard their good health by examining their family trees. The role of genetics in heart disease, high blood pressure, prostate cancer, colon cancer, skin cancer, obesity, diabetes, arthritis, depression, Alzheimer's disease, and alcoholism is discussed.

Go-ahead on altered crops is met by a go slow. Keith Schneider. *New York Times* A14 My 25 '94

(May 18) The Clinton Administration's plan to give the agricultural biotechnology industry broad authority to market most genetically engineered crops without intensive Government review has alarmed some scientists. Dr. Gus A. de Zoeten, chairman of the botany and plant pathology department at Michigan State University, and Dr. Richard Allison, a plant virologist at the university, argue that crops engineered to be resistant to viral diseases may be more risky than the government believes and therefore should not be exempt from Federal oversight.

Splashing in the gene pool. Jerry Adler. *Newsweek* 119:71 Mr 9 '92

Genetic engineering has found a home in the grocery store. In hopes of spurring the development of the industry, the White House has announced that regulations will be eased on genetically engineered products. Genetic engineering is an improvement over old fashioned cross breeding because it allows the direct manipulation of specific genes. Monsanto has introduced a gene into cotton and corn that makes the leaves of these plants toxic to caterpillars. The California firm Calgene has developed a tomato with a gene that turns off the production of an enzyme that causes rotting. DNA Plant Technologies of New Jersey has a miniature pepper on the market that is the size and shape of a jalapeno but has a mild, sweet flavor. DNA Plant also has introduced a canola oil that does not break down when heated. Genetic engineering has been applied to farm animals as well as plants, but efforts in this field have yet to prove their commercial value.

When DNA isn't destiny. Sharon Begley. *Newsweek* 122:53+ D 6 '93

Part of a cover story on breast cancer and heredity. Geneticists, who have located about 700 genes for inherited diseases, are now trying to determine why some people with such genes become sick and others do not. Some, like the gene for xeroderma pigmentosum, only become deadly in the presence of some environmental factor. Others, including the gene linked to Huntington's disease, gradually become more dangerous with each generation as the repetitions of a defective gene segment multiply. Still others, including the genes for psoriasis, diabetes, and some forms of mental retardation, produce different results depending on whether the mutation comes from the father or the mother. In some instances, including some cases of cystic fibrosis, 2 genes may work in concert.

The blossoming of biotechnology. Mark Fischetti. *Omni* 15:68–72+ N '92

The biotechnology industry has found it unexpectedly difficult to turn laboratory successes into products. Even after a new technology is shown to be effective, numerous ethical and legal issues remain to be resolved. At present, research teams are focusing on the areas of disease therapy and diagnostics, agriculture, and the creation of animals that produce cheap pharmaceuticals in their blood and milk. The Food and Drug Administration has approved more than 20 genetically engineered medical substances for sale, and more than 30 therapeutics are expected to be in human trials by the end of 1992. Nearly all these products simply treat diseases rather than cure them, however, and some of the treatments are only marginally effective. An overview of current research in gene therapy and of some of the ethical issues associated with such research is provided.

This "tree" can save your life. Sue Browder. *Reader's Digest* 142:66–8+ Mr '93

Since research indicates that there is a genetic component to nearly all ailments, tracing one's family health history can be vitally important in preventing disease. Tips on investigating family medical history are provided.

Genetic linkage: interpreting lod scores. Neil Risch. *Science* 255:803–4 F 14 '92

The development of sophisticated molecular genetic tools and the discoveries of the genes identified with some inherited disorders have intensified interest in human genetics. When the biochemical or physiologic basis for a genetic disease is not known, a sequential, random search of all human chromosomes is required to identify the mutant gene. Linkage analysis, the statistical technique used to conduct such a search, is based on the association between the frequency with which 2 genes are inherited together and the proximity of their loci. The validity of the resulting lod score depends on the assumption that the trait in question is governed by a single gene that is inherited in the standard Mendelian manner. This assumption is true of some hereditary diseases, but others involve several genes. In such cases, it is important to interpret the linkage evidence in the context of prior studies and to consider the plausibility of the genetic model used.

Biotechnology reaches beyond the high-tech West. Anne Simon Moffat. *Science* 255:919 F 21 '92

At the recent annual meeting of the American Association for the Advancement of Science, biotechnology proponents predicted that the products of the biotech industry will soon improve agriculture in developing countries. If the genomes of economically important species can be mapped, plant and animal breeders will be able to determine quickly whether a genetic cross has produced a desired trait. The writer discusses plans to improve the productivity of African N'Dama cattle while retaining their resistance in crop plants by engineering them to express viral coat proteins or bacterial toxins.

Genome research in Europe. W. F. Bodmer. *Science* 256:480–1 Ap 24 '92

Part of a special section on science in Europe. Europe must develop a coordinated genome program that can be an effective partner in the worldwide Human Genome Project. Too large for any one laboratory, funding agency, or country to sensibly undertake on its own, the Human Genome Project provides an opportunity for many scientific groups around the world to participate. Europe lags behind the United States in overall activity devoted to the project, but it is making a respectable and effective contribution. If Euope is to remain at the forefront of disease control and other areas of cultural and economic importance, each European country should have its own genome project that would both exploit genome analysis nationally and serve as a basis for international collaboration.

Molecular advances in diseases. *Science* 256:717, 766–813 My 8 '92

A special section highlights recent research on hereditary diseases and therapies that may correct them. Advances in molecular biology have enabled scientists to identify the genes that cause a number of inherited illnesses, but these advances raise ethical and social questions. Articles discuss the coming generation of biotechnology products, the use of plants as a source of biotechnology products with medical applications, methods for altering specific genes in animals to created disease models, and research into cystic fibrosis, epidermolysis bullosa, retinitis pigmentosa, Alzheimer's disease, Gaucher disease, and malignant hyperthermia.

Pigs as protein factories. Richard Stone. *Science* 257:1213 Ag 28 '92

Researchers in several countries are investigating ways to induce mammals to produce valuable proteins in their milk. Experiments are under way on swine that have been genetically engineered to produce a blood clotting protein, cattle that can make the antimicrobial protein lactoferrin, sheep with human alpha 1-antitrypsin in their milk, and goats that can produce the transmembrane protein that is defective in cystic fibrosis patients. According to attendees at a session of the ninth International Biotechnology Symposium in Crystal City, Virginia, most of the remaining obstacles are legal rather than scientific. Investigators are concerned about patent rights to the key processes involved and about guidelines being developed by the Food and Drug Administration.

Regulation of transgenic plants. Charles J. Arntzen. *Science* 257:1327 S 4 '92

The U.S. Department of Agriculture (USDA) should revise its regulations governing the evaluation of plants modified by recombinant DNA. Nearly

600 field tests of genetically engineered plants are in progress or have been completed, and the results of these studies show that such plants do not create new agricultural pests. Despite these reassuring findings, the USDA Animal and Plant Health Inspection Service continues to regulate each transgenic plant on the basis of the scientific protocol used to create it. The USDA should adopt an approach closer to that of the Food and Drug Administration, which ruled in May 1992 that genetically engineered plants should be subject to the usual levels of government scrutiny unless they possess characteristics that raise food safety questions.

Familial Alzheimer's linked to chromosome 14 gene. Jean L. Marx. *Science* 258:550 O 23 '92

With the help of chromosomal markers supplied by the Human Genome Project, researchers looking for the gene defect that causes Alzheimer's have made important progress. Gerard Schellenberg of the University of Washington School of Medicine in Seattle and colleagues have found a genetic defect on chromosome 14 that is linked to an inherited, early-developing form of Alzheimer's. Other research has found defects on chromosomes 21 and 19. There remains at least 1 other form of hereditary Alzheimer's whose gene defect has not been tracked down. The new discoveries could shed light on the important debate over the role of beta-amyloid deposition, which may either cause or result from the onset of Alzheimer's. Discussed are the direction of future research and the possible role of chromosome 14 in the synthesis or processing of amyloid precursor proteins.

"DNA-based diagnostics: the molecular medicine era is reality." C. Thomas Caskey. *Science* 258:14–15 N 20 '92

Part of a special issue on the 1992 Science Innovation meeting. Thanks to improved accuracy and lower costs, the use of DNA-based diagnosis for heritable diseases is becoming widely accepted. DNA-based mutation detection methods are available for such heritable diseases as sickle cell disease, alpha-1-antitrypsin deficiency, Tay Sachs disease, Gaucher disease, and cystic fibrosis. While laboratory operations possibly represent the simplest aspect of health care for genetic disease, physician and patient understanding and appropriate use of genetic data represent more significant challenges. Several detection methods and scanning techniques are briefly noted.

Statistical evaluation of DNA fingerprinting. B. Devlin, Neil Risch and Kathryn Roeder. *Science* 259:748–9+ F 5 '93

The National Research Council's recent report on DNA fingerprinting is based on erroneous assumptions, and the further studies it recommends will confuse and prolong the debate on this forensic technique. The NRC report accepts the argument of R. C. Lewontin and D. L. Hartl that heterogeneity among subpopulations of a large racial or ethnic group could influence the allele frequencies within each subgroup and thus invalidate the standard assumptions of population genetics. This position is a minority view, however. Most population geneticists agree that variation among individuals within the same population accounts for most genetic diversity and that major ethnic group membership contributes more diversity than membership in a subgroup. The NRC panel's proposal that 100 individuals from each of 15 to 20 populations be studied to set a ceiling probability would introduce a high likelihood of error because of the small sample sizes.

Frontiers in biotechnology. *Science* 260:875 906–44 My 14 '93

A cover story examines the current state of the art in biotechnology and the obstacles that remain to be overcome before biologically based therapies find clinical applications. The discovery that restriction endonucleases can be used to cut DNA in specific places is enabling medical researchers to increase their understanding of the human body and to begin designing therapies that intervene in specific biological processes. Articles discuss molecular therapies for cardiovascular diseases, the disruption of signal transduction pathways, the technical problems associated with gene therapy, the design of artificial substances for tissue or organ replacements, improvements in in vitro fertilization, cell adhesion and the development of anti-inflammatory agents, and financial problems that some biotech companies have encountered.

Plant biotechnology in China. Zhangliang Chen and Hongya Gu. *Science* 262:377–8 O 15 '93

Part of a cover story on science in Asia. In order to feed its population of 1.2 billion, China is employing the methods of plant biotechnology and molecular biology to improve agricultural yields. With government support, Chinese laboratories have become Asian leaders in areas such as tissue, cell, and protoplast culture. China is still struggling to catch up with the developed countries in basic molecular biology research and in policies for the safe management of transgenic organisms. The writer discusses Chinese scientists' attempts to develop pathogen-resistant plants, increase rice yield, sequence the rice genome, augment the efficiency of the nitrogen-fixing soil bacterium Rhizobium, protect crop plants from heavy metal contamination on polluted farmland, and improve the nutritional quality of various crops.

Serious flaws in the horizontal approach to biotechnology risk. Henry I. Miller and Douglas Gunary. *Science* 262:1500–1 D 3 '93

General scientific principles, particularly those derived from man's understanding of the biological world and evolutionary biology, should serve as a guide to public policies governing the new biotechnology, including those that pertain to health and safety regulation. International groups and professional organizations who have investigated assumptions about risk assessment of the new genetic modification techniques generally believe that risk is primarily a function of the characteristics of a product—whether it is inert or a living organism—rather than the use of genetic modification. The scientific organizations involved in risk assessment; the flaws and pitfalls of risk assessment experiments; the fallacy and danger of a "horizontal approach" to risk assessment; and 2 correct approaches to risk assessment are discussed.

Debating the use of transgenic predators. Billy Goodman. *Science* 262:1507 D 3 '93

In November, about 30 ecologists, entomologists, and molecular biologists gathered in Gainesville, Florida, for a workshop on the use of transgenic insects. Their aim was to draft scientific guidelines that researchers could use to help them plan experimental releases of transgenic arthropods. The potential uses of transgenic mites and insects for controlling crop pests are many and varied, but there are concerns. One problem is arthropods' mobility. There is also concern that these creatures will shift their appetites and their genes. Insects, for example, carry a wide variety of transposable elements, so-called jumping genes, that are likely vectors for incorporating foreign genes into arthropods. If the transposable element held a gene that conferred resistance to an insecticide, such a transfer to a pest would be calamitous. The genetic engineering research of biological control specialist Marjorie Hoy of the University of Florida is briefly discussed.

Genetic damages. Karen Fitzgerald. *The Sciences* 32:7+ N/D '92

Warnings and fears about potential abuses associated with genetic testing are becoming realities within the insurance industry. Consumer advocates and others have long been cautioning that genetic information gleaned through simple modern techniques or through the Human Genome Project could be used for discrimination. A pilot study published in the March 1992 issue of the American Journal of Human Genetics addressed 41 people who said they had experienced discrimination because of genetic-test results. The study showed that 32 of the cases involved unfair treatment by insurance companies. Insurance industry representatives have argued that it is fair to use genetic-test results when a condition might require extra medical care, noting that genetic predictors are similar to other health predictors currently in use. Consumer groups, however, want legal remedies to protect employees and insurance recipients from the discriminatory use of genetic testing.

Tracking down killer genes. J. Madeleine Nash. *Time* 136:11–12 S 17 '90

In an interview, genetic scientist Francis Collins discusses how tracing the origins of diseases through genetics is changing the face of medicine. Although he believes that widespread genetic screening can be beneficial, he is affronted by the idea of using the resulting information too broadly for fetal selection; it is crucial to make distinctions between life-threatening diseases and traits. Collins maintains that drawing the line in other situations, such as with genetically influenced manic-depressive illness, is difficult. So that such issues can be addressed ethically, Collins believes that people who have the most experience in philosophical and religious realms must learn about the scientific facts of genetic research.

Seeking a godlike power. Leon Jaroff. *Time* 140:58 Fall '92

Part of a special section on life in the 21st century. By the next century, knowledge obtained from the $3 billion Human Genome Project could have a huge impact on society. Scientists are working to isolate and identify the more than 100,000 genes located in the human genome, which is the strand of DNA in the nucleus of each of the body's cells, and to sequence the 3 billion chemical code letters in it. Biologist Leroy Hood of the California Institute of Technology predicts that with this information, scientists will be able to take blood samples from a newborn infant and use them as "DNA fingerprints" to learn if the child is predisposed to certain illnesses. Contentious issues will arise, however. The article discusses concerns over genetic privacy, prenatal detection of genetic diseases, and the rise of genetically engineered foods.

Happy birthday, double helix. Leon Jaroff. *Time* 141:56–9 Mr 15 '93

At a recent symposium at the Cold Spring Harbor Laboratory on Long Island, researchers celebrated the 40th anniversary of the discovery of the double-helix structure of deoxyribonucleic acid (DNA). The groundbreaking discovery was made by Francis Crick and James Watson, who were then at Cambridge University in England. Determining DNA's structure led to a Nobel Prize for Watson, who is now the director of the Cold Spring Laboratory, and for Crick, who now studies the brain at California's Salk Institute for Biological Studies. At the symposium, speakers discussed research that has resulted from the 1953 discovery, including cracking the genetic code, describing the machinery of the living cell, identifying and locating specific genes, and learning to transfer them from one organism to another. Recent and ongoing gene therapy research is discussed. In an interview, Watson and Crick discuss their discovery and their recent work.

The genetic revolution. *Time* 143:46–57 Ja 17 '94

A cover story discusses genetic research. Scientists are scrambling to unravel the mystery surrounding deoxyribonucleic acid, the spiral-staircase shaped molecule found in the nucleus of cells. Genetic engineers are decoding the molecular secrets of life and attempting to use that knowledge to reverse the natural course of disease. Even as the initial benefits of the genetic revolution start to become apparent, though, people are beginning to wonder what those benefits will cost. The dilemma currently posed by the genetic revolution involves whether people want to know about genetic defects that can't be corrected yet. Articles discuss the search for the gene that causes breast cancer; the Human Genome Project and its leader, Dr. Francis Collins; and Dr. W. French Anderson, who pioneered the first successful human gene-therapy trials.

Fried gene tomatoes. Philip Elmer-Dewitt. *Time* v143 p 54–5 My 30 '94

The U.S. Food and Drug Administration has endorsed as safe the first genetically altered food that will be sold to consumers. The product is a tomato called the Flavr Savr, produced by Calgene of Davis, California. According to the company, the Flavr Savr, which is said to offer "summertime taste" all year long, is about to make its debut in selected supermarkets in California and the Midwest. The tomato is expected to be available across the rest of the U.S. by the end of 1994. The biotech industry has hailed the FDA's approval as a breakthrough that will pave the way for approval of other genetically engineered food products, such as chickens that grow faster on less feed, snap peas that stay sweeter longer, and bell peppers with fewer seeds and a longer shelf life. Critics of biotechnology, however, continue to campaign against genetically engineered food products.

The age of genes. Shannon Brownlee and Joanne Silberner. *U.S. News & World Report* 111:64–6+ N 4 '91

A cover story examines the great strides that have been made in genetics. Although the study of heredity began in 1865 when Gregor Mendel experimented with peas, most of the medical breakthroughs in the field have occurred in recent decades. The most exciting new field is genetic therapy, in which doctors treat disease by giving patients new genes. Despite all the good that molecular medicine will do, genetic studies are also triggering a growing number of ethical dilemmas. One concern is that health insurers, employers, and the government could unfairly discriminate against people on the basis of their genes. Ethicists are also worried that as genetic research advances, parents will be able to endow their children not only with healthy genes but also with genes for height, good balance, and intelligence. Related articles discuss the important role that rodents play in genetic research and how gene therapy is leading to a cure for cystic fibrosis.

NIH, the modern wonder. Bernadine P. Healy. *Vital Speeches of the Day* 58:662–6 Ag 15 '92

National Institutes of Health (NIH) director Bernadine Healy addresses the City Club Forum in Cleveland: The biological revolution, spearheaded by the NIH, is transforming medicine and influencing the economy. The revolution involves the ability to define, recombine, and express the genes of virtually all living organisms. The NIH has begun research on gene therapy, in which a faulty piece of genetic information is replaced or suppressed. The NIH's Human Genome Project, meanwhile, is working to crack the human genetic code. Such projects raise a variety of social, ethical, and legal issues. In particular, the policy of seeking patent protection for gene sequences is an area of broad societal concern. Still, the NIH must walk these new paths to link civilizations by extending and enriching human life.

Genetic counseling. Jane Ferrell. *Vogue* 182:150+ F '92

The specialty of medical genetics is now an important part of obstetrics programs at most of America's large medical centers. With many women having children later in life, sessions with genetic counselors are becoming more common. Genetic counseling is also growing in popularity because scientists are discovering genes that are linked to birth defects. With these findings, women can be tested to see if they are at risk of giving birth to an infant who might be affected by any of a variety of genetic disorders. Still, tests are only available for about 500 types of genetic disorders, and there are reputedly 5,000 disorders in humans that are caused by genes. Advice for those considering genetic counseling is provided.

Save your life. Carol Stevens. *Washingtonian* 28:80–6+ D '92

A growing number of medical diagnostic tests are available that can help save lives. Some of the new tests are complicated and expensive, but others are so simple and inexpensive that they can be used to screen large groups of people who neither show symptoms nor have family histories of a particular disease. The fastest-growing field of diagnostic tests involves genetics. By analyzing cellular material from a blood sample, a stool sample, or a single cell of a fertilized egg, doctors can diagnose and sometimes treat certain illnesses. The writer discusses the ethical issues raised by such tests and describes new tests to detect cancers, Alzheimer's disease, immuno-suppressing viruses, Lyme disease, heart disease, hypertension, osteoporosis, bulimia, cystic fibrosis, Marfan's syndrome, hemophilia, Tay-Sachs disease, sickle-cell anemia, and certain forms of mental retardation.

The double-edged helix. Joel L. Swerdlow. *The Wilson Quarterly* 16:60–7 Spr '92

Advances in genetic research confront individuals, physicians, scientists, and society with tantalizing promises as well as disturbing dilemmas. Researchers have launched an ambitious effort to map and sequence the entire human genome. The project is still in its early stages, but practical applications of genetic research have already entered the medical field. "Biotech" drugs have spawned a $12 billion industry, and doctors have performed the first sanctioned gene therapy on a human subject. In addition, early achievements in gene mapping have led to presymptomatic screening tests for certain genetic diseases. Unfortunately, science is now at an awkward halfway point. In the case of genetic tests, for example, scientists are much better able to predict diseases than to cure or treat them. Nevertheless, there is no reason to stop or slow genetic research. In many cases, more knowledge will prove to be the best way to eliminate today's moral dilemmas.

Are we the sum of our genes? Howard L. Kaye. *The Wilson Quarterly* 16:77–84 Spr '92

Many of the scientists involved in the Human Genome Project believe that their work will uncover the biological basis for all human behavioral, mental, and moral traits. This view is misguided and dangerous. No matter how many genetic correlates of character are discovered, no biological fact can properly be said to have "logical consequences" for human nature or culture. The moral and intellectual qualities that characterize human beings have long since taken on applications and ends that transcend or even contradict the narrowly biological. Scientists and science writers probably mean no harm when they speak of the genome as "the formula for life," but such misleadingly reductive statements could subtly encourage people to stop seeing themselves as moral beings and start thinking of themselves as essentially biological beings.

Adventures in the gene trade. David N. Leff. *World Monitor* 5:34 O '92

Two occasions in which biotechnology as applied to plant genetics encountered unexpected obstacles are recounted—the creation of Frostban, a gene-altered microbe that prevents ice formation on certain fruits and vegetables down to 27 [degrees] F; and nitrogen fixation, a process in which soil bacteria capture free nitrogen from the atmosphere and join it to other elements to make protein.

Unraveling genes to find new cures. Klaus Thews. *World Press Review* 39:22–4 Je '92

An article excerpted from Stern of Hamburg. Biotechnology is paving the way for a healthier future. Experts estimate that at most, one-third of all disorders can now be cured or brought under control through medication. Drugs can only alleviate the symptoms of the remaining diseases. Modern genetic technology, however, has been used to find an easier and cheaper way to locate a cell's DNA, which stores the genetic information for the construction and maintenance of the human body. Scientists have also used genetic technology to understand a group of proteins called receptors, which protrude from the walls of cells and serve as "docking stations" for chemical substances. Gene regulators and gene therapy are discussed.

The brave new world of genetic research. Thierry Damerval. *World Press Review* 39:22–3 Ag '92

An article excerpted from Le Monde Diplomatique of Paris. Ongoing research into human genetics represents a major advance for medicine, but it also raises the question of how the knowledge gained should be applied. Genetic analysis, for example, makes it possible to detect abnormalities well before birth. Consequently, medical expenses could be predicted and could play a factor in a mother's decision to abort the pregnancy. Genetic analysis is also opening up new possibilities for employment recruiting offices, for insurance companies, and for any industry in which an in-depth knowledge of an individual is needed. The article discusses the possibility that society might eventually be divided into healthy and weak people, the media's continued focus on the positive aspects of genetic research at the expense of the complexity of the issues that it raises, and the need for citizens to be better informed about the potential consequences of genetic research.